SPRING

AN ANNUAL OF
ARCHETYPAL PSYCHOLOGY
AND
JUNGIAN THOUGHT

1986

SPRING PUBLICATIONS
P.O. Box 222069
Dallas, Texas 75222

CONTENTS

II. Pathologizing

III. Current and Controversial

IV. Jungiana

ACKNOWLEDGMENTS

To Princeton University Press for quotations from the *Collected Works (CW)* of C.G. Jung (Bollingen Series XX), translated by R.F.C. Hull, edited by H. Read, M. Fordham, G. Adler, and Wm. McGuire, and published in Great Britain by Routledge and Kegan Paul, London. Other quotations have been acknowledged throughout in appropriate notes and references.

Spring 1986 is edited by James Hillman, assisted by Mary Helen Sullivan.

Printed in the United States of America

THE ORIENTATION AND WORLD-RELATION OF ANIMALS

ADOLF PORTMANN

Biological research has treated the concept of inwardness as if it were a diseased limb to be cut off; hence that concept has been separated from the main body of science by a ligature both artificial and extremely effective. To the extent that biological research wishes to be recognized as a natural science, that excluded subjectivity must be retrieved. After that it must, with extreme care, be elevated to the status of an object of genuine research. This expansion of biology may be thought of as one of the most meaningful phases of its latest development, firmly installed next to the outstanding results of developmental physiology and genetics.

According to the interpretation of life forms especially widespread today, the organization of sensation and nerves, as well as the capacity for locomotion, is considered primarily as an instrument serving the principal function of all living things in the interest of survival. These animal functions are easy to name—finding food, locating sexual partners, avoiding enemies. In a word, animals establish just those relationships to inanimate and animate surroundings which are necessary for survival in the broadest sense. According to this view, the central nervous system is an organiza-

---

''Orientierung und Weltbeziehung von Tieren—neue Erkenntnisse biologischer Forschung'' appeared first in *Universitas* 12 (1965): 1287–300 and is printed here in English for the first time. Other papers by Portmann, the world-renowned biologist, Professor at the University of Basel, and for more than thirty years a leading figure at the annual Eranos Conferences, have appeared in *Spring 1976* and *1982*. The translator, Richard B. Carter, a Descartes scholar, is Visiting Professor of Philosophy at the University of Maryland. During 1973–74 he worked with Portmann in Basel and began the major task of translating Portmann's essays, occasioning correspondence and further meetings in Maine shortly before Portmann's death. Carter's main interest in Portmann's work lies in its challenge to Cartesian thought to expand its boundaries so as to encompass the ideas of living form and qualified evolution.

tion haphazardly emerging out of the materials provided by random living constructs in the course of indeterminate processes of mutation and selection following strict natural laws; it is an organization which came to be in this manner which accomplishes this task.

The warbler's species-characteristic singing reminds us, forcefully indeed, of another ordering of vital characteristics which differs from that in which survival holds the highest rank.[1] What we have in mind here is a set of stages *at the summit* of which we find the animal's relation to the world being effected through its inwardness and, tied up with this, we find its self-representation within the realm of appearance.[2]

To the extent that it creates an intensification of self-representation and at the same time makes for an intensification of the manifestation of inwardness, the singing which characterizes a species is like play: during singing the meaningless flow of "dead time" is transformed into meaningfully lived time; "mere time" becomes fulfilled time—that is, becomes time given form.

We know from our own experience that any mere lapse of "physical" time is difficult to bear, that it is "dreadfully dull" to us. In just this way we know that the happy life is a timeless state where the passage of time is unobtrusive. Such a frame of mind is blissful. Time then "flies"; time is then forgotten. That in dreary moments we wish to kill time, to banish it, shows what a severe and unequivocal assessment of the situation is expressed even in our figures of speech. Expressions such as "time on my hands" and "whiling away the hours" indicate very typical human situations involving individual problems which are social problems as well. The negative shading given the enduring of empty lapses of time and the positive ones granted the forgetfulness of time are both so elementary that it is perhaps not improper for us to attempt to discuss such experiences in the life of higher animals as well.

The activity of animals in zoos testifies to the necessity to give a qualitative form to time. It seems to me all animal play must be viewed within the perspective of this sense of the "formation of time." (Of course we think this way somewhat even now, but certainly not enough.) The functionless singing which takes place among immature birds and the frequent "mockery" and imitation of sounds among birds which occurs in the autumn—all this must be understood as the fulfillment of genuine free time. Even warblers imitate other songbirds. We might also do well to take note of the striking rituals displayed in higher animal life, and to discuss them both in terms of their mechanical origins and with an ear to their acoustical effects—again within this context of "qualification" of the passage of time. All these rituals obviously figure in the sphere of survival functions, and inflexible roles in fixed, critical life-games can be assigned to the effects of these rituals. According to the prevalent interpretation of behavioral research, even these "functions" have an aspect which is functionless. Manifestations such as play or species-characteristic singing are free expressions of this second aspect, expressions which lack the component of "role." It seems to me that this functionless aspect belongs to a higher level of living Being, to a primary givenness which is *served by* survival functions.

The particular case to which we may now turn is the migrating bird. To be sure we can only gain limited insight into the many phenomena presented to the biologist by bird migration or to the questions to which these give rise. A broader inquiry would demand that we examine the complicated structure of hormonal glands which provide a partial basis for the regularity of the event, in conjunction with which the phenomenon occurs. We would be forced to look into the organization of the metabolism which makes it possible for particular species of plovers to fly nonstop over more than 3,000 kilometers of open sea. Here, however, we only intend to consider the orientation of migration, that is, methods of ascertaining direction which the bird employs and which integrate its journeys into the entirety of the life of its species in a meaningful way.

By way of beginning, a richly portentous sign points our thinking in a new direction. If the sky is covered with clouds, starlings can be seen to be in a state of restless motion, flying in every direction of the compass. However, when the sun is shining a distinct choice of direction begins to show itself. The sun must be the guidepost. But how? The migration of birds does, to be sure, proceed in a relatively constant course in the same direction as long as the sun describes its daily arc. But how can these birds know the way when the sun should certainly indicate an entirely different direction in the morning from its place in the evening? We humans solve the problem by calculation; that is, we observe the path of the sun, and clock time gives us a second known quantity with which we can then determine direction.

Here it is that research has given us a plethora of results which at first sight are nearly unbelievable—namely, that a bird's brain is capable of performing the appropriate calculations for navigation, that it performs these calculations continuously and with astonishing precision, and that the bird directs its flight according to these calculations. Starlings and other birds can be successfully trained toward a definite point of the compass so that, for example, the bird always looks (say) south for its food. We cannot take time here to discuss the particulars of the experiments which lead to this and other discoveries to be discussed later on.

But a start can be made as follows: a starling is offered a circular arrangement of twelve feed boxes each of which resembles the other so as to be interchangeable, with one of the boxes (which alone contains food) always facing directly south. In four or five days the bird learns to seek out this box. Let us assume that the bird has learned that, at nine o'clock A.M. the feed box stands at an angle of forty-five degrees to the right of the sun. If it is merely this angle which registers in the bird's memory, we could assume safely that at three P.M. the bird should again seek the box at forty-five degrees to the right of the sun, now in the compass direction "west." However, the bird seeks the box at forty-five degrees to the *left* of the sun. An inner adjustment knows how to "convert" the apparent course of the sun, making it possible for the bird to adhere to a fixed point of the compass. A way to determine time accurately must exist within the inwardness of the bird. This widely varied experiment shows us that, at any position of the sun, the starling finds the direction decided on for its training and that the bird is hence truly capable of adjusting to the sun's daily movements. Once trained toward a certain direction, the bird retains its training, even without practice, for quite a long period of time. A well-trained starling performs its work in the direction to which it was originally trained for up to ten months, often even longer.

The bird's inner clock can be reset by artificially disguising its day. Let us imagine that the starling's environment is manipulated in such a way that an artificial sun rises six hours later than the sun normally would (the transposition requires several days to effect smoothly). Then, for our bird, it is three o'clock in the afternoon precisely at nine A.M. our time. Now at nine A.M. our time the bird looks south for its food at approximately forty-five degrees to the right of the sun. And it is in this direction that the bird does in fact look in our experiment; the bird is consequently directed away from the *artificial* south by a total of ninety degrees, because the time-scale required for the bird to grasp direction has been falsified by our interference.

Though we could easily accumulate particulars of this kind, we are at present only concerned with one detail: the diurnally migrating bird is equipped with a navigation apparatus which allows it continuously to determine the azimuth of the sun—that is, the particular point on the horizon *above which* the sun is located at any given time—which therefore ensures the constancy of its flight throughout its journey. We have reason to believe that this system functions just as involuntarily as the bird's sense of time—a sense which we humans also command. Furthermore, there seems to be reason to think that the altitude of the sun has little or no influence on the bird's ability to determine direction on the horizon—though there is some question as to the influence the sun's altitude might have on the ability carrier pigeons, for instance, have to "home" from completely alien surroundings.

The bird is not only a creature of light and eyes; it is also incomprehensibly adapted to the dynamics of the solar system of which our earth is a member. Today it does not seem quite so amazing to us since we know that species of crabs, as well as bees, are also capable of this orientation.[3] What is more, we know that these tiny brains can orient to polarized light when the sun is not visible. Nevertheless, we would do well to give full sway once again to our amazement in order to experience this mystery in its grandeur—an experience allowed us only by the most exact investigation into nature.

When we are struck by how very much the bird is a sun-being; by how very much the length of day in the course of a year influences the bird's hormonal rhythm; by how the sun's course governs the bird's waking and sleeping patterns—their flight will appear to us a sort of crowning of a flyer's way of being. The bird then appears to us as the most beautiful of images for such splendid day-living.

Yet we must also note the number of these creatures of light which, during the great migration, suddenly become beings of the dark. Their restlessness is the restlessness of nocturnal migration; they travel their species-characteristic routes in night-long flights and rest during the day. The general transformation of a creature which usually goes to sleep at dusk and wakes at dawn is itself mysterious, though apparently hereditarily determined.

The peculiarity of nocturnal migration has been explored intensively only in recent years. Important studies have been conducted in the United States, taking advantage of the opportunity to observe nocturnal migration as birds move across the disk of the moon.[4] Significant advances were made in Europe using radar to record nocturnal migrations.[5] However, concerning the nature of an orientation toward nocturnal migration, Dr. Sauer's experiments with warblers are the most worthwhile.[6] The special advantage of Dr. Sauer's observations is that he was able to perform particular and differing experiments using warblers which he had reared from the moment of birth in complete isolation, without contact of any sort with members of their own species.

Observation of these warblers testifies to the fact that birds isolated in this manner lapse into the restlessness which precedes migration at the time characteristic of their species, and hence that no social "contagion" is required to bring this about. This restlessness lent itself to experiments involving a cage which was shifted to a different location every night to preclude any familiarity with the locale. From the bottom of this cage which measured about four feet (ninety centimeters) at the base, the only part of the night sky which could be seen in the autumn is that where the Milky Way is located, where the constellations Cassiopeia, Lyra, and the Swan dominate the heavens (hence no other orientation points can be seen). When the direction indicated by their motions in the cage was observed, garden warblers as well as hooded warblers were seen to display, without exception, a distinct preference for one direction—south-southwest by southwest—precisely the migration direction characteristic of this species in September and October. In the spring of 1955 an experiment conducted during the return-migration phase was indicative: on three distinct nights, a hooded warbler chiefly chose the direction characteristic of spring migration, north-northeast by northeast.

The starling migrates by day. How it must question itself about those signs in the daylight sky which give it its bearings! Experiments with warblers revealed that, when the moon shone during their night-time migrations, the chart of their mapped flight directions deviated from the characteristic migration direction, that even falling stars were noticed by them, and that this caused deviation in the direction charts. Dark nights caused by cloudy skies resulted in tentative flights in all directions of the compass and a temporary spell of restlessness. Experiments with polarized light show orientation toward all possible directions, thus no special utilization of this type of light.

There is only one set of conditions under which warblers consistently orient themselves exactly toward their innately given direction of migration—when the sky is starry and moonless. Even if, as a result of artificial conditions in the laboratory, they have never before been able to see a starry sky, they choose as a direction of migration the one which is typical of their species.

The warbler's ability to orient itself during migration by means of the starry heavens must be inborn. Someone hearing of these findings for the first time will think this a very bold conclusion, an unreasonable demand on the impartial observer; but these findings do, in fact, allow us to conclude that the picture of the starry sky with its constellations contains the directing factors for orientation. Even confined to the dark, the bird obviously remains an eye-creature which orients itself to the light, to forms.

An animal with an inborn picture of the starry sky! Who, even in defense of the idea of an archetypal foundation for a world-relationship, would have allowed his imagination the luxury of such speculation! To be sure, evidence has been sought and even the earth's magnetism has been postulated, and then eliminated, as a possible indicator of direction—though, indeed, unknown physical components of the earth's force-field may yield the kind of orientation we find in night-flying warblers. The possibility of decisive experimental control was first realized through the use of the Zeiss company's planetarium. In a planetarium, the heavens can be presented in widely varying aspects to the eye of the migration-restless warbler; we can vary the orientation of the bird's visual field without altering any postulated force-field. Given that choice of direction is still the central datum, the distinct positions of this artificial heaven can only have meaning as directors.

If the results of earlier experiments convinced researchers that the warblers orient themselves by means of the starry sky, the same researchers must have also expected what the experiment in the planetarium revealed: the warblers orient themselves to this artificial night-sky exactly as they do in the open air; that is, it is the constellations which control the nocturnal migrations of these small travelers. In the spring of the year, the warblers displayed the restlessness associated with migration chiefly in the direction of north-northeast; in the fall, under the artificial starry sky, birds chose a southerly direction—and the typical migration directions for chats (*Klappergrasmucken*) and hooded warblers (*Monchsgrasmücken*) are known to be southeast and southwest, respectively. When, however, the dome of the planetarium was starless, the birds were just as disoriented as they are when the open sky is cloudy; they displayed the restlessness associated with migration by dispersing in all directions.

This artificial celestial dome gives us yet another important experimental tool. Using it we can present to the eye of the songbird a number of distinct views of the heavens, each of which belongs to a different geographical location. To begin with, an attempt was made to present the warblers with a series of pictures of the starry sky as it would appear to the nocturnally migrating bird when seen from gradually increasing southern latitudes. As long as the pictures were of the heavens at fifty, forty-five, and forty degrees of northern latitude, the birds' flight-chart remained constant; the preferred direction of migration was southwest by south-southwest. Under the natural conditions to be found at thirty-five, twenty-five, and twenty degrees north, and therefore in the area of the Mediterranean Sea zone, very frequent preference was shown for southern flight. When we artificially locate the bird beneath the view of the sky as it naturally presents itself at fifteen and ten degrees, and hence as it normally appears to the bird south of the Mediterranean Sea zone, the complete transformation toward a southerly course is made. Though the bird does not at all change position, the changing picture of the starry sky is able to produce a state in the bird's inner being correspondent to its natural behavior during autumnal nocturnal migration. Beneath the appropriate position of the stars, even when the bird remains in a planetarium itself located in the northern hemisphere, the bird changes its course in a way which corresponds to the migration course of the chat in the eastern zone of the Mediterranean Sea.

Another experiment will deeply impress us with the static power of these views of the sky. In it we suddenly shine a starry sky over the warbler under the artificial dome in such a way that the aspect of the artificial heaven corresponds to a location further west or further east. If the view on the dome is in fact the focus of the bird's orientation, it will have to feel itself to be "taken off course," certainly a contingency the bird can meet in nature when, say, it meets with strong crosswinds. If this "blowing off course" is directed toward the east, the warbler resolutely puts itself on a westerly course and hence compensates for the change at once. The more the picture of the stars in the planetarium approximates the normal time at that locale, the more a turn in the compensating western course appears, to the south-southwest. Finally, when the normal aspect of the heavens is restored, the chart of changes of intension shows the chat's proper southwestern course for fall migration.

From the preceding experiments, it is certain that the nocturnally migrating bird orients itself through its view of the starry sky; that the location of the constellations and of the Milky Way must be the variable factors; that the bird knows the way; and that built into the bird's inner being is a well-prepared structure for measuring time and for converting the position of the stars to the bird's own use. Orientation in wide expanses of darkness is carried out solely by means of the view of the stars, independent of landmarks. We are not certain of the extent to which the height of the sun plays a special part in day migration and its sun orientation, where the deviation from the meridian, the azimuth, is the decisive factor. However, we do know that, in nocturnal migration, not only the azimuth but the height of the stars is utilized by the bird in its survey of the heavens.

We know nothing of the consciousness of the tiny warbler. Hence, to claim to have something to say about the "impression" made by the changing pictures of the sky is indeed a futile game. In this instance, the inwardness of the bird is presented to us as a whole—though we are certainly able to declare that there exists in the bird, as installed by its heredity, a complicated relationship with the starry heavens; the inwardness of these feathered creatures is structured from the beginning in an experimentally demonstrable relationship with the bird's view of the night sky, in which constellations—its picture-like aspects—are the effective elements. The extent to which one and the same species of bird makes use of solar as well as astral orientation, and even of other means of direction-finding, has yet to be clarified. We are here in the presence of new insights which begin to alter our picture of living things.

As with the discovery of the sun-compass, the discovery of the bird's navigation by the stars is both a result of and a challenge to research. We are speaking here of apparatuses, of mechanisms, of descriptive paraphrases for yet unknown facets about which we at present know only that what we have discovered is a prospectively prepared, hereditarily given structure in the bird for the building of a rich world-relationship. At this time, no one can quite imagine the concrete structure which is at work here. In order to do so, we must first learn to imagine our cybernetic machines—which can perform corresponding operations—as semiliquid, macromolecular forms of being. It is precisely these machines of biochemistry and biophysics which are technically still strange to us. The search for the technology of life in the area of plasmatic events is indeed one of the greatest tasks of biological work.

However, a second task is just as important—and how different it is from the first! This is the integration of the knowledge about migration-orientation into the whole of the bird's life. To accomplish this, we must attempt to see that phenomenon in its all-encompassing reality. For a while, we must give up our perspective as inhabitants of the northern hemisphere of our earth when viewing these birds' travels. Aside from this, biology is in many cases centered on phenomena of the northern temperate zone, which of course introduces certain "standards." Others may be found through a study of tropical life—or perhaps not found at all until a worldwide comparison of very different zones can be made.

When one considers this all-encompassing relation in its entirety—and it is in this relation that the bird's form of being must be seen and interpreted—one finds a warmblooded animal, capable of powerful flight, an animal stronger than all mammals capable of flight and incomparably superior to insects in strength and aptitude for flight (however much some insect migrations might amaze or, as in the case of the locust, appall us). Everything essential to the bird's aptitude for flight strikes us as being in a hereditary, set relationship: the structure of the wings, the lungs, the tracheal sacks, the voice organs, the formation of the feathers—whose origin is still so inconceivable—not to mention eyes and hormonal system. At the head of this list we find the central nervous system which, above all else, strikes us in this way.

Here it seems appropriate to point out that not one single member of these systems is today seen as a particular something which has completely developed into what it is—despite our unshakable certainty that these members *have* indeed developed into what they are. We ought, therefore, desist from believing that we truly understand these members when, at the same time, we consider them as *continuing* to develop. The belief in this unspecified development requires as a foundation the portrayal of the already existing, that is, of the life-form insofar as we are able to discern it *here and now.*

The migrating bird's coordination to the stars and to the sun; its integration into a dynamic order which works in an unknown way with their positions; its entrance into a dynamic order as a member which includes a measure of duration to accomplish an orientation within a worldwide space to a degree which we, at this time, know of in no other animal. . . . An earlier time would more clearly recognize that all this signifies a unity of form and achievement. The order of form, function, and world-relationship lies ready in the animal as a prepared field of possibilities which await actualization as the structural dispositions of a seed await their unfolding. The potential of astral and solar guidance answers to the organs of flight and the sense of distance, that is, to the capability for a command of space of which any particular species of bird only transforms and realizes one of many sectors into a species characteristic.

This explanation may be valid only for the higher levels of organization among birds. Further research must indicate the degree of commonality of the potentialities observed among warblers. Several facts speak in favor of these potentialities being more widespread. There can be no doubt that very simple forms of life have at their command a solar determination of direction which is carried on by means of a "calculation" of the sun's position. Hence, it is not improper for us to credit the very high cerebral organization of the bird with similar potential, even if not otherwise manifest. Furthermore, migration is found even in birds having a brain structure simpler than that of the warbler.

So it seems to me more correct to attempt the investigation of bird life within the framework of an overall view which sees the bird's world-relationship as provided with a very broad field of potentialities, in particular with a rich store of possibilities for orientation. Proceeding under such a hypothesis, it will be easier for us to understand that, in closely related species, we find kinds of animals which migrate and kinds which do not.

When we see the bird's field of potential as furnished with the possibility of freedom of movement over the entire planet, the different phenotypes are not seen as levels which ascend out of the inadequate endowments found at a lower stage up to higher forms of migration-orientation and distance-ranges. Rather, they are seen as different actualizations in a system of potentials which allows for a great many variants, and in which relatively poor utilization of orientation occurs alongside richer actualizations. The way of life of the Antarctic sea swallow and of certain stormy petrels and plovers shows us the highest development into the living reality of worldwide migratory flight.

It seems to me crucial to understand that there exists from species to species a different real development of a field of potentialities given readymade to the bird in its very way of being. This is not the point at which we should give final shape to the questions which this interpretation leads us to ask, speaking as we do from the perspective of zoology. Our observations are solely intended to suggest a horizon of understanding within which biological work can be done. We are attempting to make evident, within the framework of biological research, one of the most extreme possibilities of this interpretation; we have decisively settled upon world-relation, effected through inwardness *(innerlichkeit)* and self-representation (*selbst-darstellung*) as the highest characteristic of the living being. And it is biological experiment itself which shows us how truly hidden that inwardness is—so much so indeed that our comprehension of that inwardness also includes hereditarily prepared relationships to cosmic reality—a combination of relationships which strikes us as in its way monstrous.

The enormous reach of the inherently prepared relationship reminds us of the corresponding enormity of this whole alien way of being and of the mystery at the origins of this particular inwardness. Perhaps the very scope of this prospect leads us to confront the mystery of appearance in its profundity. That mystery lies in the degree to which the living form, its coloration, its song, all of its modes of expression exceed the demands for preservation of the individual and the species. Of course, what appears is of constant service for self-preservation as well; but just as the planetary organization of inwardness exceeds the elementary achievement of preservation, so also does self-representation in appearance rise above all mere functions of preservation.

The attempt to integrate navigational ability into the whole of the feathered creature's way of seeing also casts light on the peculiarity of our human form of existence. Biological research is compelled to reflect anew on anthropology. Someone who learns of the amazing inborn abilities of the bird may perhaps regret for a moment that these possibilities have not been given to us too—perhaps this wishful thinking might go further and paint a picture of all that a brain of a complexity comparable to ours would be able to grasp by means of a hereditarily prepared world-relationship—perhaps potentialities for a relation extending far beyond the earth's orbit to other planets and galaxies. One might even allow oneself to imagine a human-like animal able to utilize unconsciously such a means of space orientation. But what would it be? An animal which, perhaps, would be safe in a complex way because of its orientation abilities—but an animal nonetheless. What is more, an animal without that same freedom of movement which birds have, showing thereby that the bird's navigation system is precisely that one fitted for life in the aerial domain. And with this it appears we have reached the limits of our comparison. It is certain that the preliminary animal stage of humans and of the higher mammals originated as already prepared from an archaic form which was very different, in its implicit world-relationship, from the bird's form of life.

Hominization is the coming-into-being of a new way of being whose peculiarity is made more obvious precisely by the otherness of forms of life such as birds display. Migratory birds supply us with insight into an ordering of living things which particularly impressively displays a rich, hereditarily secured world-relationship through integration into global events. Biological research shows us an extrahuman life whose integration into the world corresponds to a worldview similar to that used by humans for orientation in the archaic, almost Ptolemaic, system. This system, reduced to a rational statement, is that the apparent diurnal and nocturnal paths of the sun and constellations, as seen from a seemingly stationary earth, are the regulators of the course of the day and the year.

It is entirely clear that what comprises our humanly Ptolemaic world is in the case of birds founded above all in the unconscious. The warbler-Ptolemaics need not construct their worldview by any faculty of understanding, nor do they acquire their ability to navigate as we do, through maxims learned and performed in the compass of individual lives. The bird does not require work of an imaginative and theoretic order to form its world.

Despite this fundamental difference, a significant correspondence to the bird's manner of orientation does exist in the *archaic* human: there exists an originality and immediacy of worldview proceeding from the senses on which the orientation of both man and birds is based. Our spirit finds its first nourishment, its first protections in the cosmos, its earliest certainties of belief in a worldview whose sensible structures are those which (however differently we evaluate the individuals belonging to the world of higher animals from the way we evaluate individual humans) we share with the higher animals and which in the end have led to our own worldview. In this archaic way of life, there are unconscious forces of world-mastery at work, even in us today, and though they probably work quite differently in birds, their effects are not to be discounted on those grounds. Even now in the worlds conceived by the later monotheistic religions, this mastery of the task of existence is to be sought in an ordering of the senses in which unknown and known elements of the worldview harmonize into a general view which is completely Ptolemaic. But while the warbler-Ptolemaics hold firmly to their worldview cosmos, human existence is realized in the task of breaking out of it, of overcoming the Ptolemaicism of the mind. In vehement wars against tradition, in difficult defeats but also in high triumphs, the archaic worldview which we have shared with the higher animals has been definitively abandoned.

This rupture which leads us into the Copernican worldview—and over and above this, into a new astrocosmic vision—leads to strong inner tensions. Yet the world of the senses and of naive perception remains powerful in us—united as it is with fundamental forms of expression of the primitive human system and retaining significant possibilities for artistic formation—while at the same time the spirit presses toward the new worldview.

It is not entirely inappropriate to tie more closely the relationship of the preceding to the problem of human beings—the ultimate occasion for this paper. Birds' orientation to the heavens seems to me a demonstration of extrahuman life on the highest level; it shows us what is possible for the organism in respect to a complex world-relationship on earth by way of inborn means of ordering. On the other hand, the world of warblers and starlings strongly reminds us—through its opposition to our own world—of a significant fundamental fact: man is not a forlorn, lost creature dispossessed of the sheltered existence of animals secure in their prepared orientation to the cosmos. The true contrast to the hereditary certainty of animal environment lies in that orientation which is the task of our reason, and it is given through the hereditary structural ability for *this* kind of world-mastery. In the bird, as in man, the form of life is primarily provided with a largescale, hereditarily ordered system of orientation. While the one gives forced security to these hereditary controls (found in their highest form in the case of migratory birds), the other form of life, ours, offers the freedom of choice, the remorse of error, the night of doubt, but also the joy of accomplishment.

Both forms—human and animal—have their origin in hidden sources of the living being which give form to life in its particular kinds of existence; from these sources ultimately spring the world travels of the warblers, no less than man's dreams of extraterrestrial journeys. Both are obvious mysteries. Perhaps we can benefit from such unexpected findings of scientific investigations into life through being presented with yet another possibility which is strange to us—perhaps this unexpectedness gives us the courage, the daring for a new interpretation of the human. If it succeeds in this, biology's contribution does indeed hold its proper place, that which directs our gaze toward a new journey, a navigation which uses *our* endowments.

(Translated from the German by Richard B. Carter.)

1. W. Crag, "The Song of the Wood Pewee (Myiochanes virens L.): A Study of Bird Music," *New York State Music Bulletin* 334: 1–186.

2. See also A. Portmann, "The Living Thing as Prepared Relationship" ("Das Lebendige als vorbereitete Beziehung"), *Eranos Yearbook*—24; K. Lorenz, "The Inborn Forms of Possible Experience" ("Die angeborenen Formen möglicher Erfahrung"), *Zeitschrift fur Tierpsychologie* 5 (1943).

3. F. Papi, "Experiments on the Sense of Time in Talitrus saltator (Montagu)," *Experimentia*

11 (1955); K. von Frisch, "The Polarization of the Light from the Heavens as Orienting Factor in the Dance of Bees" ("Die Polarisation des Himmelslichtes als orientierender Faktor bei den Tanzen der Bienen"), *Experimentia* 5 (1949); M. Renner, "A Transoceanic Experiment Concerning the Time Sense of the Honeybee" ("Ein Transozeanversuch zum Zeitsinn der Honigbiene"), *Die Naturwissenschaft.*

4. G. H. Lowery, "A Quantitative Study of the Nocturnal Migration of Birds," *University of Kansas Publications, Museum of Natural History* 3 (1951).

5. E. Sutter, "Radar as an Aid in the Investigation of Bird Migration" ("Radar als Hilfsmittel der Vogelzugforschung"), *Der Ornithologische Beobachter* 54.

6. F. Sauer, "Star-orientation of the nocturnally traveling Warblers (Sylvia atricapilla, borin and curruca)" ("Die Sternorientierung nachtlich ziehender Grasmücken [Sylvia atricapilla, vorin und curruca"]), *Zeitschrift f. Tierpscyhologie* 54 (1957): 29–70.

APPENDIX A

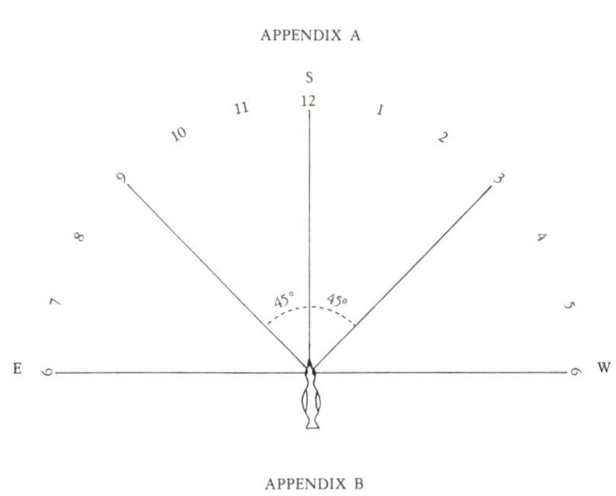

APPENDIX B

THE DISCOVERY OF THE UNCONSCIOUS IN ANCIENT EGYPT

ERIK HORNUNG
(Basel)

The mind of the ancient Near East was deeply interested in the stars. The Babylonians established the beginnings of astronomy by exact observation of the movements of heavenly bodies. In the epoch of the Old Kingdom, the Egyptians of the third millennium B.C. turned their eyes upward to that region of the sky from which the stars never disappeared but, as stars around the North Star, made their way across the horizon. The pyramids were lined up with the northern skies, and there the soul *(Ba)* of the dead king rose up to unite with these "immortal" stars. Also, statues (in which the dead were to continue to live) had their eyes directed upward seeking visual contact with this other heavenly world.

Besides being oriented to the north and up toward the sky, the Egyptians had always been oriented to the west, to an underworld kingdom of the dead into which the sun could be seen to sink every evening. In some of the very early pyramids we find both directions combined: the entrance to the step pyramid of king Djoser is directed northward; the entrance to its "South Grave" is toward the west. Thus, from the earliest royal funerary texts known to us, the Pyramid Texts of the twenty-fourth century B.C., both regions of the other world are important. The western kingdom of the dead, however, is rather negatively viewed as a place full of dangers.

But in the early Middle Kingdom, the nineteenth century B.C., a radical change takes place; henceforth, the Egyptians' interest and regard concentrate more and more on the underworld, the deep part of the world

---

Erik Hornung, Professor of Egyptology at the University of Basel, lectures regularly at the Eranos Conferences. His many publications of original research and interpretation include works on Underworld texts and Books of the Dead. His *Conceptions of God in Ancient Egypt: The One and the Many* (Cornell Univ. Press, 1982) stands unique in the field.

from which they expect rejuvenation and regeneration in life and death. Instead of the natural science of astronomy, the Egyptians develop a "science" of the underworld; and the deep part which we shall plumb here opens up for them deep insight into the world of the unconscious. Whatever ideas about the fate of the dead they developed are not open to our methods of inspection, but their insights into the depths of the human soul continue to be relevant.

How did this turning to what was inside come about? The framework of rules and regulations of the Old Kingdom, the first great flowering of Egyptian history, crumbled around 2140 B.C. owing to internal causes and not from any disturbance from outside.[1] A people that had erected mighty pyramids for eternity and monumental structures for the continuance of the pharaoh's life, a people that had created wonderful statues, reliefs, and paintings, suddenly found itself in dire poverty without the protection of a strong kingship and without the organized administration and supply system that until then had been most effective. The reaction to all this is grief and despair, and an attempt to express what was frightful in poetically formed laments. This new insight into the negative traits of human character which can be no longer suppressed is frightening. Previously, the Gods and the kingship had been held responsible for any breakdown in order, but now there is an openly expressed feeling that really the "heart," that is, human free will, is at the root of all the trouble.

Human beings alone are capable of destroying everything that has been created. In Neferti's prophecies, "not even the black under the nail remains," and the sun God Re has to undertake anew the act of creation. And the possibility of a new creation becomes an idea in aid of overcoming chaos; politically, the idea leads to the establishment of the Middle, and later the New, Kingdom; psychologically, to the insight into the necessity of continual regeneration to work against the negative tendencies of rigidity, decay, and disintegration.

The roots of this insight reach all the way back to the beginning of history. By the time of the First Dynasty, Egyptian kings were conducting a renewal festival, the so-called Sed Festival, whenever they had been successful in completing thirty years of rule.[2] They were not just celebrating a jubilee but a true renewal and rejuvenation of the king by means of cult ceremonies—the burial of the statue of the 'old king' and a ritual run of the 'new' pharaoh, who thus began a new period of rule, having regenerated his powers. For the Egyptian of antiquity, thirty years was a generation, after which a change in the royal office was necessary to enable a new generation to take over; there is some evidence that, in predynastic times, violence brought about this change before a symbolic cultic renewal was accepted.

The symbolic character of the Sed Festival is already apparent in the Old Kingdom because it was a part of the regular decoration of the royal mortuary temples. Also, it is performed for kings who have ruled for fewer than thirty years and never celebrated an actual Sed Festival. We see here the hope that these renewal festivals continue beyond death, that the departed king celebrates in the other world an endless series of Sed Festivals and, in this way, is always rejuvenated; statues and reliefs show the pharaoh in special festival costume guaranteeing the "eternality" of his celebration.

Sed Festivals are also for the Gods. In the Egyptian belief system, the Gods are, like humans, subject to aging and death, and in need of renewal and regeneration. A myth of the New Kingdom (preserved in *The Book of the Celestial Cow*) describes how the sun God Re, the creator of the world, became old and how this aging process loosed the powers of destruction and disintegration; humankind became enraged against the God. Re then destroyed much of it with his "eye" as a fire-spitting uraeus. It was only a trick of the God, again reconciled, that saved some of humankind; since then, those saved have been at war among themselves. The Celestial Cow carries the disappointed Re up to heaven. Gods and men separate. In this myth, we have the Egyptian variation of the Flood tradition—with destruction by fire instead of water.[3]

As in the breakup of the First Intermediate Period, this myth of the Celestial Cow also shows the wickedness in the hearts of humans as visibly causing loss of the harmony of the beginning. But in the world's aging process, the myth discovers a still deeper cause and shows how unavoidable is its development. Only regeneration will reverse the process and overcome the wicked. Those living in the First Intermediate Period already feel this, but they doubt this reversal is possible in the here and now. Thus the prospect of hope from a disappointing and chaotic here and now is directed to another world from which peace, justice, and happiness are expected. In his poems, the author of the "Dispute of a Man with his *Ba*" glorifies death as freedom, as a way into the "Clarity of Heaven," as a happy homecoming after a long stay in a foreign land.

Death becomes the entry gate into a world that, from this time forward, one dedicates oneself more and more to filling with life, bringing it out of the impenetrable darkness into the light of the nocturnal sun. By the time of the Middle Kingdom,[4] we find in the funerary texts detailed descriptions of individual regions of the other world with some separate illustrations. In the *Amduat,* the first Book of the Underworld of the New Kingdom,[5] word and image come together in a unity to illuminate in the greatest detail "a hidden place" (the *Amduat*'s original title is *Writings of the Hidden Place*), into which the sun descends every evening, where the dead and the sleeping also have their abode. In our dreams, then, we can meet the Gods and the dead because sleepers spend their time in Nun, in the primeval ocean from which every morning the sun rises; "Nun" in Coptic is the underworld, that underworld into which Jesus goes down to the dead. In water and darkness, the "secret" deep part of the world is concretely and externally experienced, but in the analogy of dream and death we have, in the human unconscious, an opening into that deep part of the other world. In this truly "secret" abyss, the walls between this world and the other are transparent because the speech of images removes this separation. For the Egyptians, images of nature, the behavior of certain animals and plants can become statements about fate after death. For this purpose they draw upon the huge supply of images lying ready to hand in the depths of the unconscious.

Before we descend with the sun God into that deep part, we shall have to discuss how the complex and contradictory character of human nature was regarded in pharaonic Egypt.

The Egyptian language has no word corresponding to the Greek word *psyche* or the English word "soul." Hence the most recent stage of Egyptian, Coptic of the Christian era, borrowed the Greek word. Change of religion thus becomes anthropological change. In prechristian religion, we do not find man split into body and soul. There is, rather, a much more complex concept of a difference between conscious and unconscious contents of the soul.

In the *Ba,* represented by a bird with a man's head or less often by a ram (for *Ba* also means ram), a person is quite consciously present; should a person lose consciousness (e.g., by fainting or shock), then one's *Ba* temporarily leaves one. The *Ba* is, as is the psyche, a freely moving element that leaves the body at death. A person, however, is always seeking to be reunited with it. While the body is resting in the deep underworld, the unencumbered *Ba* roams throughout the universe even to the skies, the world of the stars; from there the *Ba* always returns to the body, sharing with it those material interests, like food, drink and sex, which the *Ba,* unlike the psyche, also needs. The power of metamorphosis—the power to appear in another being, to expand consciousness in Gods and in humans—is part of this mobility.

But the *Ka,* the other "soul" of the Egyptian, is fundamentally bound to the unconscious. With outspread arms (this is the hieroglyph of *Ka*), it protects and carries the person up from the depth. It is the source of life energy, the creating and nourishing forces which are at work in humans, and also the magical forces. Whoever disturbs appetite and the sense of well-being disturbs the *Ka,* a person's inner mental balance. Maspero saw the *Ka* as the human double, but Thomas Mann defines it more precisely as "spiritual body . . . along with the body." In persons' names, divinities are called the *Ka* of living men, and an old saying has it that the dead man "goes to his *Ka*"—he sinks into the unconscious and leaves his *Ba,* the conscious part of his soul. In the deep part of the other world, the *Ka* regenerates all creatures. It is also the force that keeps the sun in its path.

Along with the *Ba* and the *Ka,* there is yet a third concept of the soul, the *Akh* which acquires significance only when a person is dead. It is one's form of life in its totality in the other world. This totality is usually portrayed as a trinity of body, *Ba,* and shadow. The shadow, the true companion—and, for the Egyptian, a copy of the person—is supposed to return, as does the *Ba.* For the shadow's role is carrier of the energy which gives it the power of astoundingly rapid movement.

While the *Ka* determines the condition of the soul in the deep part of the other world, the heart directs thinking, feeling, willing and acting as the center of one's waking consciousness. Humans do what their hearts order them to do. Thus when one comes to be examined before the Judgment of the Dead, one's heart must be in equilibrium with the *Maat,* the correct order of the world, and it must give an accounting of its owner's activities. The heart incorporates human free will which can also do evil, and the tale of Sinuhe shows that it determines conscious and unconscious action which is not explainable by reason.[6]

Among the peculiarities of Egyptian anthropology is the great significance of the body in life after death. Existence in the other world is unthinkable without the body. From an early date, enormous efforts were made to conserve it as completely as possible. Mummification was to keep the body in the same condition it had been in life. Even the rapidly decomposing internal organs were embalmed, at least symbolically, and placed under divine protection so that, within reason, nothing should go astray but be kept for the "eternity" of the other world.

As realist, the Egyptian knows that this effort, not always successful, often is a total failure; he therefore avails himself of a number of complementary aids. Amulets representing different parts of the body protect these parts and serve, if needed, as replacements for them; further, those statues placed in the grave offer the *Ba* and the *Ka* a kind of "reserve body" of the deceased with which they can also unite themselves if the body is destroyed. Most of the statues and other images of the deceased have his name and title on them because the person's name was of his essence and can also, as a further component, represent him; one is "kept alive" by having one's name pronounced or written out. The reality that a person's name and title possess for an Egyptian expands the possibilities of one's further existence.

This further existence signifies rejuvenation and regeneration of the person's physical–psychic totality. Just as the land of Egypt is changed and rejuvenated every year by the inundation of the Nile—mythically, the primeval ocean Nun—the *Ba* of a person that goes off into the Nun's depth is fitted out with a rejuvenated body. This process includes freedom from the confinements of the mummy form (which serves only as a protective covering and not as a permanent form for the body).[7] The sensory organs are also renewed to function again. In the depths of the unconscious, consciousness is always renewed.

The conveyance which carries the Egyptian down into these depths is the sun God's barque, and the customary transport routes in the valley of the Nile are water routes. Thus both Gods and the stars use boats. The evening sun as a *Ba,* in ram-head form, enters into the underworld to give light to the dead, and in the most extreme depth, to unite them with their bodies. Re's nocturnal journey is thus a process of regeneration and of becoming whole again. C. G. Jung's individuation process has already been compared with it.[8] In his novel *Passage de Milan* (1954) Michael Butor—in an analogy full of connections for the division of night in a Parisian apartment house—has made use of the twelve hours which it takes to make this journey, undertaken every day, through the underworld.

The schedule of twelve nocturnal hours gives the Egyptians the opportunity of dividing the way into the depth and of clearly delineating the completely different regions Re passes through. Other means of giving form to this journey are the gates which interrupt the voyage of the nocturnal sun and block the intruder's way—guarded by demons whose names ("Bloodsucker," "Fire Sprays with his Eyes," etc.) or forms (serpent, crocodile head) indicate how dangerous they are. In older underworld books, each nocturnal hour's area is closed up within gates and separated into three sections: in the middle one, the sun barque travels with its crew; the other two sections indicate the topography of the region of both banks of the underworld stream.

Crossing over into the other world does not happen all at once. First, one must traverse an extensive intermediate region that takes all of the first nocturnal hour; the *Amduat* indicates the extent of these spaces by describing the distances in thousands of kilometers. Only after this intermediate region and the first gate are passed does the true other world begin, the essence, places, and qualities of which are now lit up by the nocturnal sun. So they become visible, temporarily raised up into the conscious before they again sink into darkness as the sun God moves on.

The sun's nocturnal journey with his entourage is a return voyage. It goes from west to east, contrary to the usual course of the stars; it goes from age to youth, from death to life. Old man sun becomes a "small child" born each morning from the Goddess of heaven, Nut. This other world's paths do not run like those in the here and now—not straight, but crooked and zigzag; things and living creatures are turned on their heads; orientations are removed.

This topsy-turvy world, full of meaning, reminds us of the conditions of the world before the creation. The sun God takes the two creating powers, Hu and Sia ("Utterance" and "Percipience"), into the depth with him to renew the work of creating the world. By means of his light and his voice, he causes the darkness and its secrets to be made visible, filling with new life the mummies stiffly lying there. These then freely, without let or hindrance, step out and answer the sun God after they have joined their *Ba*. But the topsy-turvy (other) world influences their voices; distorted, they do not sound human but like the bellowing of bulls or the screaming of falcons, like the rushing of winds, and the buzzing of bees. Only the God's ear transforms these sounds into words that can be understood.[9]

The Egyptian did not expect resurrection to happen at the end of time but every night as a renewal from the unconscious. The renewal becomes possible because the "way back" which the sun takes is also valid for the passage of time which in these areas is reversible. The *Amduat* illustrates this process in the last hour of the night: the sun God and his followers enter, "weak with age," "grey-haired," and "ancient," into the tail of a monster serpent, proceed along its backbone, and "come out as rejuvenates" from its mouth, becoming newborn children. The other direction, from tail to mouth, makes clear the reversal of the time axis, and in the books of the underworld the serpent is the phenomenon of time made visible. The serpent's straight or sometimes curled body signifies the time continuum from which discrete time quanta are "born" as stars or in feminine forms, are again "swallowed," finally losing themselves in eternity.

Rejuvenation of the sun and all people occurs in the serpent's body. Yet the most dangerous enemy of Gods and humans, called Apophis by the Egyptians, has the same form. Apophis is darkness blocking the sun's light, the cosmic personification of evil, bringing the sun and all movement to a stop and desiring the destruction of everything that has been created. Apophis tries to suck up the whole primeval ocean, Nun, to dry up the God's waterway upon which the sun's ship sails.

This theme indicates the enormous, mysterious attraction of the deep which threatens to drag everything down and keep it there. But the God's assistants cripple the serpent, thrusting their lances through it, and burn it limb by limb, thus forcing it to yield up everything it has swallowed. Even though the accounts stress that the serpent is totally "destroyed," it is confronted anew every night because it too belongs to the process of regeneration and individuation. The double characteristic of protection and threat that the unconscious has for man is incorporated in a strikingly simple form, the *ouroboros.* The Egyptians devised this form as a symbol in the New Kingdom, passing it on to gnosis, alchemy, and mysticism—the serpent in the form of a circle biting its own tail and, among the Egyptians, enclosing the sun God who is rejuvenated in it, becoming a child again.

Apophis and the demonic guardians of the gates are not the only dangers to be confronted in this depth. Serpent-heads without number illuminate, as living torches, the darkness while also threatening the passer-by with their poisonous breaths; crocodiles leap forth upon one from all quarters. Spells and charms drive them back. Traps threaten the *Ba* in the form of a bird.[10] Many of this other world's localities are ambivalent like the "Lake of Fire," whose water is fire but, for Osiris, refreshingly cool and a source of nourishment, a "Lake of Life."

The fire's force is only turned against those whom the Judgment of the Dead has condemned, those who have shown themselves to be enemies of the cosmic order. For them the underworld is truly hell, and the Egyptians filled their religious books with visions of their punishment.[11] In these places of punishment, lying deep in the darkness where the sun will never shine, wicked dragons range, mutilating the condemned and punishing them limb by limb. They cannot achieve that unity of person for which they strive; they can only partially realize their lives and are in constant torment, threatened with the total destruction of their existence in the "Place of Destruction" in the underworld's deepest abyss. In representations of the Judgment of the Dead, there waits by the scales the figure of the "Devourer of the Dead," a creature part-lion, part-hippopotamus, and part-crocodile, that symbolizes this danger of complete extinction.

This is the Egyptian form of the vengeance of hell as we know it from the drawings and paintings of the Middle Ages; the Egyptians invented the fire-filled kettles of the Christian hell. A tomb painting from the Roman era in Egypt shows a similar kettle next to the scales and the "Devourer of the Dead";[12] schematic representations of the kettles which are held up out of the depth have already been found in the Underworld Books of the New Kingdom. Serpents spew their "fiery" poisonous breath into the seething kettles containing the "enemies" of the cosmic order. In the *Book of Gates* the "Great Fiery" appears as a giant serpent in the ninth hour of the night to set fire to the chained enemies, and in the eleventh hour of the *Amduat* they are pitched headfirst into fiery pits.[13]

Fire, most effectively and in the most visible manner, destroys the enemies of order. In the Egyptian version of the Flood, the myth of the Celestial Cow, not a flood, but the fire of the sun's eye, destroys rebellious men.[14] In the Nile valley, water was perennially a life-giving element visible in the flooding that every year rejuvenated the land and fertilized it anew. The world's depth is filled by the water of Nun, the primeval matter which, together with the darkness, preceded the creation and has since been regenerating the sun every night. A "Lake of Life," Nun rejuvenates all creatures but is not to be approached by any hostile powers.[15]

The powers of evil are held in the most extreme depth, chained and guarded, so that they may not break into consciousness and wreak evil there. We know that they are immortal, that their destruction is never definitively successful; it suffices to keep them shut up in the underworld, withholding from them any power over consciousness. They have a place in the underworld; and this makes sense because confrontation with the powers of destruction and disintegration, with the disorganized chaos before creation, and with the threat of total destruction belongs with regeneration.

Here the dark and destructive sides of the psyche are not denied but made visible. At the same time, however, they are bound up in a strict, clearly delineated order, forming the deepest layer in which vengeful and punishing fantasies are lived out. These places of destruction are, in the texts, consigned to the most extreme depth which the nocturnal sun never illumines. Only in the punishment scenes does the *Book of Caverns* omit the small disk which indicates the presence of the sun, thus stressing that there the sun does not shine but eternal darkness reigns.

We meet the Gods in a higher light-sphere, high above the abyss of destruction. The Egyptian Gods dwell in the other world. On earth their images, and especially the pharaoh, represent them. But whoever goes down into the underworld meets the Gods face to face, looks into the "Face of the Sun" which, in a scene from the *Book of Gates,* is drawn through the underworld to shine in the depth. Hundreds of forms of the Gods fill the banks between which the sun travels. The Books of the Underworld describe these forms individually or as a group.

What we notice among these divine forms are the numerous animals and the curious connection of animal heads with human bodies. Superior to a human in many ways, the animal reveals a divine power beyond itself which determines its existence. Thus the Egyptian religion uses animals to make a statement about the essence of the Gods and to deliver in visible images an idea of how differentiated and contradictory this essence can be: a lion's head on a Goddess draws attention to her wild, dangerous, and unpredictable nature, but the cat form indicates her mollified, merciful aspect; cow horns or the form of a cow, motherly support, protection, and rejuvenation of the dead. All these representation-modes can come together in the same Goddess—for example, Hathor—because only totality can circumscribe the abundance of divine forms. Thus each animal does not have to be assigned to a definite divinity. Many of them stand as a general symbol for the overcoming of death, others for the power and triumph of the pharaoh. In every case the animal is a bearer of meaning and can, in visible form, make a statement about an invisible world.[16]

During the day the sun God travels as a falcon across heaven; evenings, he enters the underworld as a ram (the sign of the *Ba*-soul) and in the morning leaves as a beetle, as a scarab which, the Egyptians thought, comes into being (as does the sun) "by itself." In between lies rejuvenation; in this interval, however, takes place the meeting of the night sun and Osiris, the ruler of the dead. Egyptian theology of the New Kingdom sees this meeting as an analog to the reunion of the *Ba* with the body—Re as *Ba* is joined to Osiris to form a unified creature speaking with *one* mouth. But every morning the connection is broken again. Osiris stays behind in the depth while Re rises up to heaven, just as the *Ba* separates from the body and goes its own way. The noctural journey of the sun to Osiris is also seen as Horus's journey into the underworld, bringing to his father help against his enemies; in the equation Re-and-Horus we find the bracket which connects the sun's course to the Osiris myth, giving the dead person a double guarantee of continuing life.

As yet, no myth of antiquity has aroused such intensive interest as that of Osiris, Isis, and Horus. In a clearly conceived tale, it brings together a multiplicity of elemental life situations and constellations, including death in an especially gruesome and drastic form.

Osiris's brother Seth murdered and dismembered him, throwing the body and limbs into the Nile, abandoning him to decomposition before Osiris had produced an heir. As a usurper Seth seized the sovereignty for which he had striven. But the exemplary love of Osiris's spouse Isis triumphs over death and injustice. Grieving sister and widow, she is not satisfied with the lamentation for the dead for Osiris; helped by her sister Nephthys and other assistants, she collects the scattered limbs and body of the God and calls them back to life, receiving from him a son and heir—"Horus arose from the seed of his father when he was already in a state of decay" (*Book of the Dead,* Spell 78).

This supernatural birth foils Seth's plan, and he now moves the old fraternal quarrel to a new plane. He seeks to keep Horus from his father's legacy by virtue of his superior strength. Isis has to protect her young son from Seth's persecution and wiles. Further, by her art of magic, she must guard the child against many other dangers until he has grown. Finally, Horus triumphs over Seth. In an old version of the myth, Seth escapes punishment, is reconciled and given part of the world—he keeps, fit for someone like him, the infertile area, the desert. Osiris, the king of the dead, now continues to live on earth through his heir; out of sorrow there is joy. The continuity of rule is assured.

In Nephthys, the "Mistress of the House," we encounter the prototype of the selflessly serving person who lives only for others; Seth, however, incorporates the other extreme, bold insolence and lack of consideration, pursuing only his own interests. One of his animals is the hippopotamus, both powerful and fat. But this God's true animal is an animal of legend defying all zoological definition. He thus demonstrates Seth's uncanny nature which cannot be categorized, whose empire is the desert and foreign places.[17] Seth is not evil; he is the challenge which every orderly system needs so that it does not become rigid; thus this destructive activity finally leads to renewal and regeneration of order. Not he, but Horus who represents justice and legitimacy, retains lasting sovereignty.

During the sun's nocturnal voyage, we meet all the persons of the Osiris myth again. Here Seth appears under the control of the powers of order as the sun's helper against the Apophis serpent into which he thrusts his spear. Isis helps him with her magical powers, but she also repeats the lamentation for the dead Osiris and the jubilation for the revivified God.

Isis's lament and jubilation never cease, for they encompass the heights and depths of human existence. In daring visions the Egyptians have illuminated the inner space of the soul to the limits of nonexistence. They have not chosen it as a resting place but have perceived it as a regenerating through-passage, as a source of daily renewal. This successful integration of the unconscious gave them the mental balance and cheerful humanity whose effect to this day, beams out from the paintings, statues, and written testaments of the Egyptians.

(Translated by A. K. Donoghue)

1. For the historical evolution, cf. A. H. Gardiner, *Egypt of the Pharaohs* (Oxford, 1961).

2. C. J. Bleeker, *Egyptian Festivals* (Leyden: E. J. Brill, 1967), pp. 91–123.

3. E. Hornung, *Der ägyptische Mythos von der Himmelskuh* (Fribourg, Switzerland: Universitätsverlag, 1982).

4. R. O. Faulkner, *The Ancient Egyptian Coffin Texts* (Warminster: Aris & Phillips, 1973–78).

5. E. Hornung, *Das Amduat, die Schrift des verborgenen Raumes* (Wiesbaden: O. Harrassowitz, 1963–67).

6. M. Lichtheim, *Ancient Egyptian Literature,* vol. 1 (Berkeley: University of California Press, 1973), pp. 222–35.

7. E. Hornung, "Vom Sinn der Mumifizierung," *Die Welt des Orients* 14 (1983): 167–75.

8. I. Clarus, *Du stirbst, damit du lebst: Ägyptische Mythologie in tiefenpsychologischer Sicht* (Fellbach: A. Bonz, 1980).

9. *Amduat* (see n. 5), eighth hour.

10. T. G. Allen, *The Book of the Dead or Going Forth by Day* (Chicago: University of Chicago Press, 1974), chap. 32 and 153.

11. E. Hornung, *Altägyptische Höllenvorstellungen* (Berlin: Akademie-Verlag, 1968).

12. F. W. von Bissing, *Annales du Service des Antiquités de l'Égypte* 50 (1950), pl. I after p. 576.

13. Texts and representations: E. Hornung, *Ägyptische Unterweltsbücher,* 2d ed. (Zürich and Munich: Artemis Verlag, 1984).

14. See n. 3.

15. *Book of Gates,* sixteenth scene (n. 13).

16. E. Hornung, *Conceptions of God in Ancient Egypt: The One and the Many,* trans. J. Baines (Ithaca, N.Y.: Cornell University Press, 1982).

17. H. te Velde, *Seth, God of Confusion* (Leyden: E. J. Brill, 1967); E. Hornung, *Eranos Yearbook* 46—1977, pp. 429–32.

NOTES ON WHITE SUPREMACY
Essaying an Archetypal Account of Historical Events

JAMES HILLMAN
(Thompson, Connecticut)

My point with these pages is rather simple, but we shall traverse a tight and sinuous course getting there. Even going fast, it will take a while. So that we know where we are heading, I'll summarize before setting out.

Our culture, by which I mean the imagination, beliefs, enactments and values collectively and unconsciously shared by Northern Europeans and Americans, is white supremacist. Inescapably white supremacist, in that superiority of whiteness is affirmed by our major texts and is fundamental to our linguistic roots, and thus our perceptual structures. We tend to see white as first, as best, as most embracing, and define it in superior terms.

Moreover, *horribile dictu,* I shall present evidence showing a similar supremacy of white in black African beliefs, enactments and values, which suggests that dilemmas in society attributed to ethnic bigotry have sources that are fundamentally difficult to modify. "Fundamentally difficult" because the fantasy of supremacy, I hope to show, is archetypally inherent in whiteness itself. If this can be shown, I would also be offering a paradigmatic example of archetypal priority in human affairs. This implies that human affairs may be affected by an archetypal psychologizing which is at the same time a therapy, in this case a therapy of racism at its structural roots. So the immediate purpose of these remarks is to differentiate this archetypal—that is, geographically distributed, temporally enduring and emotionally charged—structure inescapably intrinsic to whiteness,

---

A shortened version of these notes was read at the annual meeting of the Inter-Regional Society of Jungian Analysts, Birmingham, Alabama, October 1985. The notes emerged from the author's double concern with alchemical psychology on the one hand and political psychology on the other.

and indicate a psychological mode of ameliorating the archetypal curse of supremacy beyond the usual and necessary societal measures.

Status of Color Terms

In 1969 Brent Berlin and Paul Kay[1] of the Berkeley anthropology department published a universal scale of color terms that claims a differentiated evolution from paired white and black, then red, followed by green/yellow or yellow/green, blue, through brown to purple, pink, orange and gray—eleven in all. Their hypothesis is twofold. First, all languages follow this scale, that is, if they have only four terms these will be white, black, red, and green-or-yellow, and if a fifth and sixth, these will follow the universal scheme, e.g., yellow (or green) and blue as sixth. Second, this scale projects an evolution in color linguistics from primitive few to sophisticated many color terms. Their research with ninety-eight languages claims to have established semantic universals, or archetypals, of color terms.

Since 1969 anthropologists and linguists have attacked this thesis. There are counter-instances[2] from Samoan, Uto-Aztec, Hopi, Vietnamese, etc., regarding the order of color terms in the scale. That white/black always comes first and is always present has not been controverted.

Berlin and Kay present black/white as opposed. That the findings regarding white/black *as opposites* have not been contested indicates something important about the unconscious universality in "white" anthropology of a perspective which, when it sees white and black, sees them as opposed. Later on in these notes we will examine the relation between oppositional thinking and whiteness. Now our concern is only with the placing of white in this worldwide survey: that it comes first, is virtually universal (all languages have a term for it), and that it is paired with black.

Three Meanings of Whiteness

The Indo-European root of *white* means "to be bright," and brightness as shining, lustrous serves to cover many of the reported meanings of white in the Berlin–Kay study. For instance, the *Oxford English Dictionary*'s first definition of white states:

> Of the colour of snow or milk; having that colour produced by reflection, transmission, or emission of all kinds of light in the proportion in which

> they exist in the complete visible spectrum, without sensible absorption, being thus fully luminous and devoid of any distinctive hue.

White, however, also means blank, pallid, ashen, colorless. The Greek *leukos* describes snow, silver, silver/gold, gray dust, pale yellow, white hair, fair skin, sun, light, aether, sheen of glassy water, clear voice, happy day.[3] Latin distinguishes *candidus* (our candid, candidate), "a glistening, dazzling white," from *albus,* "dead white, not shining."[4]

To escape the philological and etymological traps of "root" and "true" meaning, we use an archetypal approach, distinguishing three main constellations in our cultural tradition as if each were a perspective governed by an archetype. The assumption here is that how white is seen and spoken about express fundamentals of the psyche.

A. The first of these is the whiteness of heaven or white as perceived through the archetypal perspective of the heavenly, spiritual father, familiar to us mainly in Christian contexts.

This "heavenly" convention appears established already in ancient Greece: Priam sacrifices a white ram to the sun. Hera wears a headdress white as the sun, and in the tragedies we find phrases for white-winged birds of dawn and radiant white-horsed day.[5] The common Greek phrase *leuke 'eméra,* "a good day," means literally a bright, happy, white day.[6] Such are the days high on Olympos where the Gods dwell (*Od.* 6 : 4445). Plato says, "White is the color most proper for the Gods" (*Laws* 12 : 956a). A priest in a fragment from Euripides says (fr. 472 : 16–19): "Wearing white robes / I shun the generation of mortals / and not touching a coffin / I have kept myself from eating the flesh of animals." White means ritual purity and spiritual sanctity, admitting neither the blackness of death nor the redness of flesh. For Aristotle (791a) the air, if pure, is *leukos*. Air takes on color only when sullied by matter. The constellation in Aristotle's thought of white, air and purity on the one hand and matter as negative, mere potential and the locus of decay on the other yields that narcissistic definition of God that stalks our western centuries like a white ghost—God whose activity is immortality (286a), who cannot be injured or wronged (109b), is better than humans (116b) utterly self-sufficient (1245b, 1249b), needing no friendship (1244b, 1159a), that *actus purus* by which God is defined also in St. Thomas and Leibniz, "fully luminous and devoid of any distinctive hue," as the *OED* says of white.

Christian texts continue this first sense of white, though we find it in Daniel 11—"to purge and make white"—and in Psalm 51—"Wash me and I shall be whiter than snow." The apogee appears in the Apocalypse in Revelations, a book shot through with whiteness, chapter on chapter—clouds, horses, angels, elders, thrones—as the transfigured Jesus in the Gospels is white (which makes some wonder what color he may have been in his quotidian form). As white and wheat are cognate, the wheaten wafer commutes into the whitened body of the God.

The heavenly sense of white extends beyond our western-northern Greco-Judeo-Christian conventions. Dominique Zahan, the major expert writing in French on Central and West African peoples, states that white for Black Africa—the Bambara, Hausa, Baranga and others—refers to the world beyond. Human beings descend from heaven where all things are white, which accounts for the idea that black babies are supposedly white for a few seconds after birth. According to Zahan, an albino in Black Africa is blessed by celestial lightning, initiates are smeared with white indicating their union with the divine, and abnormal creatures are blanched in boiling water to prepare for their return to heaven. Peace, order, joy, abundance, the ancestors, home—these are all characterized by white.[7]

Victor Turner lists twenty-three meanings of white in Ndembu. These include: goodness, strength, purity; free of misfortune, of death, of sorrow, of ridicule; ancestors; laughter; fertility; to sweep clean, wash; to make visible or reveal.[8]

> . . . whiteness, more than any other color, represents the divinity as essence and source, as well as sustenation. Whiteness as light streaming forth from the divinity has, in the sense we are considering here, a quality of trustworthiness and veracity. . . . White is also unsullied and unpolluted. . . . Behind the symbolism of whiteness, then, lie the notions of harmony, continuity, purity, the manifest, the public, the appropriate, and the legitimate.[9]

"An Ashanti priest will . . . dress in white as if death does not touch him."[10] Among the Nilotic Mandari, "White is the beneficial colour, used to express preferred moral and intellectual qualities, to represent high status. . . ."[11]

That Black Africa could be converted to the white Christ, accept salvation through a white redeemer—accept the very idea of needing salvation from sinful darkness to eternal light—seems given as much by the supremacy of white as by supposed superiority of theological doctrine. Conversion, we might say, was archetypally prepared. For the Cherokee, white represents "peace, happiness and the south";[12] for the Navaho "natural goodness";[13] in Madagascar "white: light, hope, joy, purity";[14]

on the Malay peninsula "white is generally an auspicious color."[15] In the ancient magico-religious system embodied in Hindu philosophy, the three *gunas* that enter into all things are black *(tamas),* red *(rajas)* and white *(sattva)* which is the subtlest, the primal breath-energy *(prana),* the mind or consciousness, as well as "peace," "stainless," "serene clarity," "illuminated repose, benignity, and understanding."[16]

Such notions as these may have affected the perceptions of indigenous peoples worldwide when, for instance, it was reported to Moctezuma, "Our lord and king, it is true that strange people have come to the shores of the great sea. . . . They have very light skin, much lighter than ours." Possibly says Frederick Turner, Moctezuma "was thinking that perhaps these strange white beings from the east were the returning god Quetzalcoatl. . . ."[17] What has happened to the dark animal suspicion that belongs to the encounter with anything strange? Does whiteness suppress instinctual awareness, banning from consciousness the very possibility of harm with white peace and white love and white unconsciousness, as the Navahos use white clay to drive away enemy spirits? "On the Atlantic coasts of Africa, the European ship met neither resistance nor surveillance."[18]

> They saw a great boat appear on the wide sea. This boat had white wings flashing like knives. White men came out of the water and spoke words no one could understand.[19]

This particular arrival was met with fear and "flights of arrows," for white signified the other world beyond, and its fearful ghosts.

If conversion was archetypally prepared,[20] the conquests of Mexico and Peru, the colonialization of India and the Indies by a handful of "white" men (that is, Christian Europeans), the collapse of Native American societies and of so many different peoples of so many hues, continuing even to the cargo cults of the twentieth century, these genocidal and cultural catastrophes exemplify then not only the victory of linear history over cyclical myth, of technology over ritual, of monotheistic abstract order over polytheistic concrete life, of capitalism over barter, of belief in impalpable transcendence over animistic and aesthetic immanence, of alcohol over botanicals, of superego over id—then, in addition to all these reasons for ruin, we shall suppose an underlying archetypal account which colors all others, the a priori supremacy of whiteness.

B. The second group of meanings connotes innocence. We forsake the archetypal father in heaven for the archetypal child: guileless, simple, without pollution, corruption, illness or evil. White as a lamb; harmless as a little white lie; pure as driven snow; lilywhite; the whiteheaded boy and white son who are the best, the favorites.

In contemporary psychological jargon, this sense of white supremacy appears as the innocent unwoundedness of "primary narcissism," a condition that imagines cure as an *imitatio Christi:* wounding, bleeding, descent the inevitable course of treatment, the analytical bitter sop, the soldier with his pike now analyst pointing out, penetrating insights.

Supremacy here connotes the beginning in heaven or elsewhere, spotless, before the Fall, prior to birth by blood to earth, the supremacy of perfection in the bright Garden, child of God, knowing all things save its own nakedness.

So alchemy gives white names to the *materia prima* or first condition. I have excerpted these from the more than one hundred terms for the *materia prima* in Ruland's dictionary of 1612: white moisture, white smoke, white magnesia. Tin, on account of its whiteness. Bride or Eve. Milk of Virgin; pure, uncontaminated virgin. Heaven, for it is light and bright. Lye, for it washes and cleans the metals. Urine of the white calf. Marble, crystal, glass—which are clear and intelligible. Spring. Moon. Garden. Silver. Whiteness. The Lamb.

These appear at the beginning of the operations which sophisticate the soul, operations which macerate, putrefy and dissolve this primary innocence. White is but a blank and ignorant page, the requisite stupidity at the start.

If heavenly whiteness maintains its supremacy by including all hues and unifying them into one self-same translucence (white is the entire visible spectrum undifferentiated by sensible degrees—*OED*), this second kind of supremacy maintains purity by exclusion. Innocence excludes: "innocent" literally denotes an absence of noxiousness; without harm or hurt—a privative notion of supremacy. Its confines, like a child's, are too small for differences which then must come from elsewhere, from others like Herod, the Romans and Jews, as if its kind of supremacy invites the spoiling attack upon it. For this perspective, white and black are not merely a pair but an opposed contradictory pair. Black becomes necessary to whiteness as that co-relative by means of which white takes on its defensive, exclusive definition as im-maculate, un-polluted, in-nocent. Because its purity necessarily constellates that which its purity cannot include, it cannot move unless it falls. Both perfect and precarious, it calls for help to defend its supremacy, victim of its own exclusiveness experienced as victim of inferior others.

This kind of white supremacy appears in contemporary analytical theories that divinize an immaculate child who is led by others to its fall into sexuality; seduced, abused, perverted, abandoned. I am thinking of the innocent child archetype and its white supremacy that informs the theories of Alice Miller and Jeffrey Masson; and I am thinking of the childish idealization and disguised white supremacy of their followers, seduced by innocence and perverted from the darker paths of depth psychology by abandoning the polymorphic multicolored child for one innocuously, narcissistically white.

Because it is the effect of white, whether pigment, light or word, to unify differences, the first set of meanings—heaven, divine, spirit—and the second also lose their distinction. The notions of white derived from the child become spiritualized; childlike ignorance receives spiritual authority, the child protected by the father, innocence going about the father's business with a mission, innocence now a superiority, not a frail beginning or narcissistic blindness. Yet more, yet worse, the spirit becomes absolute, that is, absolved from the woundings of doubt, guilt, sin, shame, self-awareness. (Self-sameness hardly promotes self-awareness. Hence Jung defines individuation as differentiation, not unification.)

The adamantine spiritualization of innocence together with the arrogance of the spirit regarding its own ignorance—this doubling of the two supremes generates a consciousness numbed by its own light. It cannot feel that purity may be murderous. "The Murder of the Innocents" also means *by* the innocents, by those who are supremely innocent of their own murderousness. "White massacre" and "white terror," it has been called.[21] What this white consciousness doesn't know or feel or see can't hurt it, so it thrives on repression and idealized projections.

Here are generated those conundrums of evil in western theology (since there is no place for the non-white either in heaven or child), the racial bigotry in society, and the denial of *Geisteskrankheit* in psychiatry (the spirit cannot be sick, it has been argued, only the human mind or soul). And here is generated the negative cast to the unconscious, the very definition of the unconscious which we shall take up further on. Moreover this white supremacy is reinforced, nay promulgated in our culture, as the essential nature of God by the union of the Father with the Son. So of course our culture suffers from "white love": love both as a patriarchal command and as a childish innocence, a bland love of candid commitments, ever subject to and surprised by incursions of yellow choler, deceit and jealousy, scarlet inflammations and the black depths of hell.

C. The third convention locates white in the context of anima. Robert Graves's *The White Goddess* is the classic amplification for this theme. Jungians are familiar with the symbolic association of white with anima from Jung's alchemical writings on Luna and the *albedo* (*CW* 14), a topic I have treated in two long essays.[22] In brief, this convention equates white *(leukos, leuki)* with female, soft, vulnerable beauty. It appears in Homeric and Classic Greece and even earlier in Egyptian painting where it was customary to depict the female flesh in white. In Aristophanes' *Ecclesiazousae* women take over the city; one heroine stands all day in the sun to acquire a manly appearance (suntanned) by losing her "whiteness."

In Homer Penelope is fair, Hera is white-armed, Aphrodite shows white arms, and the Maenads white limbs and white feet. Helen, Arete, Andromache, Nausicaa are all white-armed, and for Sophocles Antigone is white. For Euripides Medea's neck, Iphigenia's neck, Phaedra's neck, Alcestis's flesh, Glauce's foot and neck—all are fair, pale, white.[23] (Shakespeare's Juliet, Rosalind, Marina, Helen, Helena, Hermione and Rosalind each have white hands, Ophelia a white bosom.) Even Dionysos in the *Bacchae* (1.457)—to emphasize his softness and vulnerability—has fair skin *(leukos)* which to Pentheus means effeminacy.

Leukos as effeminacy appears also in the Greek term "white-livered" (our lily-livered) meaning cowardice.[24] To the Greeks white-livered also meant simple-minded, silly, foolish (that is, naive) emotions, since the liver is both the seat of darkblooded passions and of black bile in Greek imaginative physiology. A white-livered person, to fatten the metaphor, would have no stomach for fight (would show the white flag, the white feather, throw in the towel), no guts, no balls, thus suggesting that effeminacy or anima-dominance corresponds with a childlike naivete in regard to darker (black, blue and red) passions; no passion for darkness.

Despite the anima meanings of beauty as fair-fleshed, bright-skinned and radiant, white here no longer implies an innocent child-like supremacy. White here no longer signifies divinely immaculate, without bruise or blemish. The inferiority is retained, not cast off, because the anima sense of white is self-weakening. Whether called effeminacy, softness or vulnerability, this notion of white admits a mortal inferiority within itself. The soft white spot (*Iliad* 22 : 321–25) at Hector's throat is the place Achilles aims to strike his mortal wound. Anima white remembers disease and death.

This vulnerable whiteness parallels the whitening in alchemy. The albedo is both hard as silver and soft as down: a firm hardening of the reflexive mind, the mind a polished objective mirror without a beam of its own intentions, together with a quickly warming, gently yielding vulnerability in soul.

Supremacy—even if not in the heavenly and spiritual sense[A] or the childlike and innocent sense[B]—may nonetheless still be implied by anima whiteness.[C] For the experience of the psyche's whitening is usually one of comforting ease, a lifting of the burdensome nigredo where all was occluded. During the nigredo the psyche feels a downward pull into its heaviness and a backward attraction into the past, remembering traumas. We experience memories traumatically: we explain ourselves with materialized reductive accounts and define reality depressively as hard, unimaginative and degrading. Reflections made in a mirror darkly.

The albedo lightens the mood, as putting white around small areas of color in a painting washes the colors out, leaching away their intensity. Now one rests content on the moon, supreme in one's lustrated psychological distance, Luna the Queen, all conflicting hues faded into the equanimity of white, the psyche so beautiful, everything metaphor, so apparent, so resonant. The smug, smooth supremacy of anima-fusion. Leucothea, the white goddess, saves the distressed at sea;[25] "the association of white with deities who bring calm is related to the whiteness of calm itself."[26] The very essence of white, the shape of the atoms of which it is composed, according to the atomist theory of Democritus, is "smooth," "shadowless" and "translucent."[27] "That whiter skin of hers than snow, and smooth as monumental alabaster," says black Othello (V,2,4) overtly of Desdemona, but as well describing an idolized anima whose smoothness presents no dark faults, no rough edges.

"... the *albedo*, which for many alchemists was the climax of the work" (*CW* 14, §388), delivers one over to the smooth white notion of wholeness where, as Jung says, "all becomes one" (ibid.). The shadow of the primary white, those tinny, milky, calflike conditions, become imperceptibly blended within the smoothened lunar unity. All the while believing oneself having achieved the opus, one is actually back again at the ignorant beginnings of the prima materia. Delusions induced by supremacy. To obviate this dreamy enchantment of soul by its own lunatic potential, a lunacy not experienced as mad but as chalk cliffs in the moonlight, so various, so beautiful, so new, alchemical advice says to increase the heat toward yellow and red. *Circulatio:* move back down and around. *Iteratio:* repeat, yet again and again. For the true alchemical white bears blackness within it, the blues of memory and regret,[28] as it intends toward a further dawning, a waking-up to itself by yellowing. It is not sheer white, mere white, in fact, no longer purely white at all.

A yellowed white that is more complex than pure, stained and thus able to stain, "to tincture" as the alchemists said, rather than merely to reflect, a contagious white that goes out of itself, spreading, recalls a fourth set of meanings: white as sick. I refer especially to the Biblical Hebrew *laban* instanced in the descriptions of leprosy in Leviticus, evoked by Coleridge in *The Rime of the Ancient Mariner:* "Her lips were red, her looks were free, / Her locks were yellow as gold, / Her skin was white as leprosy." I refer also to the phlegmatic humor in ancient medicine—cold, moist, inert—still with us today in the fear of milk products and excessive mucus, and also in medical language: leukemia; "the whites"; candida; tuberculosis once called the "white plague"; and the white-skinned, white-boned, white-haired, white-spittled specter of old age. We seem now a long way from the supremacy of white with which we entered into this essay. But the conclusions are yet to come.

A Moral Conclusion

Precisely this pallid leprotic aspect leads to the moral of these notes. "Moral" because issues of shadow necessitate a moral style of rhetoric. (One sure way to recognize shadow in the room is that discourse takes on a moral tone. "Should" not only trumpets the superego, it announces the fear of shadow.) The moral of these notes concludes that *white casts its own white shadow,* has its own disease, which as white is indiscernible and especially indiscernible to consciousness defined in terms of light.

Since the albedo phase is essential to if not equivalent with the *unio mentalis,* we feel we are of one mind, spooky doubts, darkened suspicions, internal dissension dissolved. We feel cleansed of shadow in the supremacy of white. Hence some alchemists believed the work stops here.

However, as Navaho white clay drives away enemy ghosts, so the "moral aspect of whitening"[29] may in fact be that self-deception in which the ghosts are driven out by blending in, formulated in philosophy as the "identity of indiscernibles." Although the differences between white and its shadow are not perceptible (e.g., we can't see the shadow in our bright ideas, good deeds, true beliefs, honest motives, beautiful feelings or any of the other Christian virtues which ennobled the conquests, glories, triumphs and spoils of the Christian mission), they are there nonetheless. Christ and Caesar indiscernible. As Leibniz argued, intrinsic differences even if indiscernible nonetheless exist.[30] The shadow of white may be white but still shadow. The moral task now becomes not do good and be true, harmonize and integrate, but *diakrisis.* "Discern or perish."[31]

Discernment conceived by an unsophisticated fantasy of "differentiating the opposites" removes white's shadows to the black at the other end of a spectrum of extremes. As a Jungian, I always imagine that extremes touch, for this was one of Jung's favorite maxims. More than touch: mingle, intertwine, conjoin, so that all white angels are at the same moment white devils, the healing *pharmakon* of white love also a poison, and white magic or white maenadism (E. R. Dodds) no less magic or maenadic because they are white. Let us define evil not as the absence of good *(privatio boni)* but the very presence of good, in all ways and forever, inextricably coincident. "There can be no consciousness without the perception of differences," says Jung (*CW* 14, §603). Consciousness to deserve the name mustn't shirk awareness of the blinding propensity intrinsic to each particular event nor escape through the dodge of opposites which permits one side to be right and white, therefore immoral and unconscious. Consciousness is at once fundamentally clarifying *and* self-deceptive, so when white appears in dreams or thought or life be glad and beware.

All lustrous moments in therapeutic practice, whatever the brilliance, the smoothing of trouble, cleared perceptions and saving transfigurations through insight, suffer the delusional snow job of white supremacy which forces distinctions into oppositions, splitting white and black; ". . . schizophrenics showed higher preference for white" in the color pyramid test, and "the incidence of white as a pathognomic sign of schizophrenia has also been reported" by other investigators.[32]

Differences neither compete, contradict nor oppose. To be as different as night and day does not require an opposition of night and day. "White and black," says Jean Buxton of the Mandari, "are *used* [my italics] in conflict situations or where opposition is to be stated."[33] They are pragmatically opposed for a functional or aesthetic purpose, like laying a black stroke on a white field to make a strong statement. But this pragmatic use to underline conflict does not entail an ontic necessary opposition, that in their nature and being black and white are opposites.

Berlin and Kay claim white and black as language terms appear always *together.* It is Berlin and Kay who declare the terms opposed, applying their white (un-)consciousness to the linguistic material. That separation, that apartheid, forces inherent distinctions of shadow within each to become extrinsic, white losing its shadow in blackness.

Black, we may note, does not show this inclination toward supremacy. We saw that in Black Africa as reported by Zahan it is white that carries the meanings of special, heaven, good, divine, while black as reported by Victor Turner connotes disease, sorcery, bad luck, suffering, work and death—rather like our black plague, black magic, blacking out, black ball and black list. (These "inferior" meanings of black for people called black would be expanded and further differentiated were that the object of these notes.[34] Further, in painter's language, "shade"[35] distinguished from hue, tint, intensity and value means mixed with black; here too black welcomes a sense of shadow. And I am told that American black vernacular is full of shades for so-called black people: yellow, brown, dusky, coffee-colored, blue-black, even pink. We may speak in a literary style of white complexions as ruddy, pearly, tanned, peaches-and-cream or sallow, yet white collapses to the one category. Whites are not vernacularly grouped in complective subsets. (The nickname "red" refers to rufous hair and freckles.) My point is that only white has this built-in supremacy which annihilates its shades.

That whiteness does not admit shadow, that its supremacy rejects distinctions and perceives any tincture as dullness, stain, dirt or obscurity has been remarkably described in Tanizaki's little book *In Praise of Shadows.* The Japanese novelist brings out the failings and destructive soullessness of our white napery, gleaming silver and bright glassware that sterilize eating; the dazzling electrification that hardens and stiffens our interiors; the white porcelain toilets in white-tiled cubicles which make our dark excrements only more smelly, alien and repulsive; our bleached white book papers inviting no meditative calm; our white hospitals in which human illness becomes obscene.

By "inherent distinctions of shadow" within white, I mean the differentiation of innocence, for instance, into shadings—the demivierge, ignorant of biological sexuality but corrupt in charm; the young innocent, without shady intentions but white-lies aplenty; the good innocent, pure as driven snow and cold as ice, or filled with the milk of human kindness and a mouth full of pap. I think of the white lamb's wolfish neediness, of the white Goddess who is as well a demon of death, of the white noise from the White House. And I think of the white powders, white pills, white papers, white sprays, white 'dust,' white smoke, and whitewalls that appear in dreams to eradicate insects, anesthetize feeling, eliminate or hystericize sensations, justify wrongs, and establish the perimeters of our interior defensiveness. I mean also a differentiation of the heavenly. There are millions of stars, each heavenly body with its own place and grade of light, and in ancient thought endless numbers of archons, angels, spirits, each characterized with its own translucence. Why are there so many angels, and why so tiny that they can dance on the head of a pin if not to force intense concentrated noticing of their individual differences, each unlike each, differences lost when we whitewash events with simple, pallid words like good, spirit, love, child, conscious, anima, whole, heal, help—the words that carry forward the white mission of psychology. Angels, each with a name and face and voice, are crucial for distinguishing the varieties of whiteness. Without angelic distinctions, white consciousness becomes unconsciousness, that is, insane. The loss of the white angels is the fall into the white race.

"Discern or perish"—if consciousness is differentiation as Jung says, then each white will be differentiated by the image which presents it. Which *sort* of white in this dream, in that behavior: deathly? bridal? lamblike? milky? Which *shade* of white: blinding? ashen? silvery? When, where, and how does it appear: as an unwritten page, assuaging aspirin, the handle of a clipping scissors, the long dress of a dead woman? Each white bears its specific triumph and specific threat together; that is, each white offers one of the supremacies together with the shadow specific to that supremacy.

A Psycho-Historical Excursion

"White casts its own white shadow." This conclusion may be bettered to say, "white sees its own shadows in black," not because they are inherently opposed but because it is archetypally given with whiteness to imagine in oppositions. To say it again: *the supremacy of white depends on oppositional imagining.*

When the perception of white here entails simultaneous perception of black there, we have what is commonly called projection. Projection, however, works two ways. The second reverses the first: perception of black there results in identification with white here. This psychological rule became the historical case. The "white" identity of northern-western Christian men begins (in English) only after their arrival on the shores of West Africa, mid-sixteenth century.

> The most arresting characteristic of the newly discovered African was his color. Travelers rarely failed to comment upon it; indeed when describing Negroes they frequently began with complexion and then moved on to dress (or lack of it) and manners. . . . Englishmen actually described Negroes as *black*—an exaggerated term which in itself suggests the Negro's complexion had powerful impact upon their perceptions. Even the peoples of northern Africa seemed so dark that Englishmen tended to call them "black" and let further refinements go by the board.[36]

The first use of white to characterize an ethnic group occurs in 1604 *(OED)*, after the perception of Africans as black. By 1680, says Jordan, "taking the [American] colonies as a whole, a new term appeared—*white*" (p. 95). Although Negro slaves were baptized and Negroes joined the church as full members (Massachusetts) as early as 1641, the most common term the settlers used for themselves during the seventeenth century was "Christian." "Christianity had somehow become intimately and explicitly linked with 'complexion'."

> From the first, then, *vis-à-vis* the Negro the concept embedded in the term *Christian* seems to have conveyed much of the idea and feeling of *we* as against *they:* to be Christian was to be civilized rather than barbarous, English rather than African, white rather than black. (p. 94)

Oppositional thinking "from the first," as Jordan says. Thinking in black-and-white. Peoples identified *by* color become identified *with* color; Africans who did not perceive themselves as "blacks" become "blacks," a discovery of white consciousness. West African slavery begins psychologically not in the structures of black society, as hinted by Braudel,[37] nor even in the inherent supremacy of whiteness described above (**A** and **B**), but rather in the yoke of oppositional thinking. Even if calling oneself black reclaims with pride an African heritage and history of suffering, one nonetheless remains within the same archetypal construct of whiteness present in the mindset of the first traders. "Black" may be beautiful, but it is a white term. The Greeks, Hebrews, Romans, Crusaders, Venetians, Portuguese, Spaniards—none called themselves "white," none intruded into the human world of peoples the archetypal realm of color, inflating and deprecating humans with transhuman values—a mode of speech arising with modern consciousness and reaching its apogee in nineteenth-century colonialism: "red" indians as savages, the "yellow" peril, the "white" man's burden. Freedom from slavery here means freedom not merely from the white master and the mastery of whiteness; it means mastering the oppositional mind that sets white against color of any kind, the color bar, the essential constituent of whiteness.

The whitening of the West proceeded concurrently with the blackening of the rest. This blackening was stored inside the term *black* long before any English-speaking person touched the shores of West Africa.

> As described by the *Oxford English Dictionary*, the meaning of *black* before the sixteenth century included, "Deeply stained with dirt; soiled, dirty, foul. . . . Having dark or deadly purposes, malignant; pertaining to or involving death, deadly; baneful, disastrous, sinister. . . . Foul, iniquitous, atrocious, horrible, wicked. . . . Indicating disgrace, censure, liability to punishment, etc."[38]

A logic of opposites preserves white from these foul notions. White would have no deadly purposes, could not be iniquitous and wicked or liable to punishment. White folk are clean and censure-free,[39] as if white ethnic identity was itself a baptismal font assuring salvation, as if alone the language of white by virtue of its archetypal resonances could restore purity, eliminate sin and guilt and shame, as if the mixture of blood in half-caste and half-breed were the worst of all betrayals of white group identity[40] because it empirically destroyed the logic of opposites which declares that where white is there can be no colored admixture, the logic of the excluded middle lived socially. How else account for the moral unconsciousness, the incredible self-satisfaction and smooth justifications that accompanied the massacres by white Christians (despite their scriptures and ethical jurisprudence) in that holocaust prolonged through centuries our history books call the ages of discovery, expansion and the primacy of the West.

Not only did the northern-western mind uncover its white predilection while discovering West African peoples during the period 1550–1700; concurrently, it made discoveries in modern optics. Huygens, Kepler, Galileo, Descartes, Grimaldi, culminating in Newton's *Opticks* (1704—though presented in writing and lectures some thirty years earlier), which finally reduced the spectrum of hues to fractured bits of white light.

Alchemy too flourished during these same one hundred and fifty years, yet that flourishing is better regarded as one of late summer ripe for fall: a huge harvest of literature, bounteous compendia, dissemination (Ruland's dictionary in 1612 and Agnolo della Casa's eighteen volumes [1592–1618] each with nine-hundred leaves, or Borel's bibliography in 1654 with 4,000 entries). Alchemy had itself attained the *multiplicatio,* the late stage in the processes it describes, and achieved the *exaltatio* influencing courts and kings. It was due for return to the base condition of the *prima materia,* the state into which it subsequently fell (until revived by Jung), rejected as fraudulent, worthless or "only the beginning" of the rising new sciences.

The new optics embodied other fantasies about the sensate world, displacing the alchemical importance of colors as revelations of essential nature with inherent psychological properties, the very manifestations of the divine, God's rainbow, guaranteeing His presence in the raiment of things. Alchemical colors are in the world, of the world, and tell about the world. After Newton and Locke (1690, first draft 1672), colors are refractions of light, neither a sign of mystery nor an essential virtue, rather a secondary effect produced in the human understanding by abstract laws. The subjective qualities of color were relocated in the literalized subject. No color in the world of things out there, only the mechanics of light. With Newton, white assumed supreme status, the equivalent of light, primary—and colorless.

> Hail, holy Light, offspring of Heaven first-born!
> Or of th' Eternal coeternal beam
> May I express thee unblamed? Since God is light,
> And never but in unapproachèd light
> Dwelt from eternity....
>
> (*Paradise Lost,* III, 1666)

The coronation of white by the scientific paradigm removed it from the chromatic scale of alchemy. There it takes its value partly from its place within a process; there it could be loaded with inferior, vulnerable, and tangible descriptions such as we saw in the white names for the *prima materia.* "Eternal" and "unblamed" whiteness describes sheer supreme white, God and child indiscerptible, self-reflection gone. White, risen to become God of the colors, became an invisible transcendent, a fantasy most welcome to contemporary deism in religion and the emerging enlightenment in philosophy.[41] The coronation of white also suggested an easy equation: black = the deprivation of light (optics) = *privatio boni* (theology) = without the light of reason (philosophy). This equation in social history promotes missions to benighted natives, the moral justification of slavery, and the identification of dark-hued peoples with brute and bestial nature, dumb, without intelligence (for which modern terms are still borrowed from this topos: "bright," "brilliant" and "quick").

The relation between colors and light has a mythical parallel in Plutarch's *On Isis and Osiris* (probably written in Delphi ca. 120 A.D.):

> The robes of Isis are variegated in colour . . . the robe of Osiris, however, has nothing dark or variegated about it, but is of one simple colour, the colour of light; for the origin of things is unadulterated and the primal element which is spiritually intelligible is unmixed. (§77)

The reduction of iridescence to white light translates mythically into the subjugation of Lady Isis to Lord Osiris, visible things to invisible law, polytheistic variety to monotony. The brilliance of Newton's achievement with the prism demonstrates the nature of light at the expense of the light of nature, both the clarity of enlightenment and the illusion inherent in it.

The White Unconscious

The convention informing geographical discoveries and the expansion of white consciousness over Africa continue to inform psychic geography. The topological language used by Freud for "the unconscious" as a place below, different, timeless, primordial, libidinal and separated from consciousness recapitulates what white reporters centuries earlier said about West Africa. From Conrad's *Heart of Darkness* to van der Post's *Venture to the Interior*, Africa and the unconscious allegorize the other place. (Psychology conveniently imagines white men projected their unconscious onto Africa but projection works two ways; geography's Africa appears as psychology's unconscious.) "Just don't stay in the tropical colonies too long; you must reign at home," writes Freud[42] in 1911 to Jung, who himself made the African journey fourteen years later, describing the vast lands and dark peoples he encountered in language he applies as well to the immemorial unconscious psyche. Dream images such as forests and jungles, caves and oceanic deeps, cellars and swamps—any place without light, unmapped or trackless (without *methodos*) may casually be said by "depth" psychologists to symbolize the unconscious. Part of psychology's myth is that the unconscious was "discovered"[43] as its contents are "explored." Even the notion of the underworld as black rather than grayish, misty or invisible bespeaks white supremacy.

Moreover, the "discovery" of an unconscious separate from consciousness, as a black continent separated from white penetration into it, maintains the very unconsciousness within white which the idea was invented to wound. For the idea of the unconscious follows from the idea of repression which Freud thought more important than (and even the cause of) the unconscious. "The unconscious" is first of all a pragmatic idea, a moral idea, aimed to correct repression, i.e., that unconsciousness of the northern-western mind about the limits of its awareness.

Rather than imagining the unconscious as a dark region discovered by Freud, the intrepid adventurer, let us consider him the social moralist, who presented a chastening idea in the language of a new place, an idea that had become morally necessary—and abhorrent—to white supremacy, since it limits its freedom (as a *Grenzbegriff*), corrects its Christianism (as Freud and Jung differently insisted), expands its ethnocentricity to collective universals and, by coloring its whiteness with variety, challenges its superiority and stains its innocence. It is this unconscious white consciousness that is the proper object of depth psychology, depth come home to roost, out of Africa; depth in Freud's sense of the omnipotence fantasy and Jung's sense of shadow: ever-present and always mine, the very me I am now, imagining myself eternal and unblamed. And so these notes restate Jung's insistence upon shadow and his moral anger over Albert Schweitzer's African mission. Not there, here.

Yet Jung's notion of shadow, like that of the unconscious, rises within the white topos we are describing. Though serving the aim of self-correction, the ideas of shadow and unconscious maintain the theory of opposites and locate consciousness with light, day, bright, active, etc. And so the entire modern psychological effort to raise consciousness, and the ego drafted to enact the endeavor, is one more manifestation of whiteness, perpetuating the very fault it would resolve. The project can never succeed since the unconscious it would redeem lies in the instrument of its intent, in the eye of its light.

Psychology assumes repression, projection and the unconscious are bedrock laws of the mind. They are fundamentals, however, only within the topos of their origins—the white North, the Christian West, modern history. That these notions are a priori to the northwestern soul is truly descriptive of that soul: it represses, it projects, it is hugely unconscious. That is how this soul identifies its consciousness, defends it and survives. As long as white consciousness maintains itself with metaphors of optics (maximum clarity and centralized focus) and symbols of sharpness (the knife of discrimination), where can the excluded animated others appear except elsewhere as "projections," animisms, miracles, synchronicities. The unconscious becomes the vast excluded of psychoanalysis, a huge submerged iceberg, consciousness its little white pointy head.

If there is a history of consciousness, then there is a history of changes in the defense mechanisms on which this consciousness depends. As Bailey shows (below), projection indeed has its history in magic lanterns; and alchemy's notion of projection had nothing to do with our modern ideas about the unconscious psyche of the operator. Alchemy confined projection wholly to the *opus:* the throwing off or effusion of one substance upon another thereby effecting transmutation. These older notions were 'internalized' by modern psychology's fantasy of subjects throwing emotional veils over objects. Now at the century's close we recognize a post-modern experience of this same term. Projection is no longer confined to a subjective psychic activity which externalizes those fantasies incompatible with so-called consciousness. It is no longer merely a matter of the white mind laying its dark attributes onto black people.

We now experience projection to be the very essence of our high civilization; the white supremacy fantasy lies latent within its forward, upward thrust, its *wurf* or throw. Projection in every project, program, problem, prognosis, production, profession. What is not projection? By what criteria discern action from acting out? What does the white eye see on the horizon but the white heat of its own projected extinction? Projection now is not of the repressed; it refers to what is actually lived in our most deliberated projects. What we call consciousness today is simply projection. Even the one whose project is to take back projections, to integrate them in a wider whole, is also a "who" embraced by the same projective topos, a figure in the cave. The shadows dancing on the wall cannot be seized by the centered attention of more focused light, taken back as mine; they may be caught only fleetingly, peripherally, by a glance, a moment askance that respects their independence like a flicker off a rock and given with the rock. For no longer do we want to tell the dancer from the dance or be the squatters on the cave floor trying so sincerely not to project. "Who" is to say what is conscious, what projection, or whether the rockwall is more unconscious than the watching we. I turn my face to the wall; the wall is what I have.[44]

Jung's *later* articulation of "the unconscious" coming from alchemy starts in the rockwall, in nature, not in the human mind. This "unconscious" is a dark field of small perceptions (Leibniz), of living sparks, scintillae, fishes' eyes—*lumen naturae,* the light of nature as islands of consciousness. This "unconscious" offers a radical shift of metaphor with ethnic implications. As there is always and essentially shadow within light, so there is always and essentially light within shadow. Neither "conscious" nor "unconscious" can ever be literally such; in fact, "discern or perish" becomes impossible since that by which we discern itself requires discerning. There is no standpoint that is not unconscious. There is no light all white, no immaculate perception. We are all mulattoes of the mind.

And all things from beasts to buildings, phosphorescent, darkness glowing; and so all things showing blackness, whether imagined morally, racially or geographically, would flash their fishes' eyes; brightness streaming from nearer than the firmaments of heaven, now closely, panluminously, in the sticks and stones that break white bones. "The unconscious" becomes again a destructively wounding self-description required for self-regulation, even self-preservation, limiting whiteness not by a dark surrounding opposed but now by the uneasiness occasioned by alternative lights. Fish everywhere, everything fishy. No longer the topological Africa of the mind down there; instead opacity, suspicion, irony, complexity, density and some quiet joy at being relieved of the white man's burden of clarity.

Alchemy and Society

The alchemical appreciation of color recognizes shadows within any hue. The *rubedo* or reddening, for instance, may be the goal, equivalent to gold, the elixir, tincture and philosopher's stone. As well red may be violently sulphuric, or a mercurial phase, or the primal crude earth of Adam. That is, red may describe conditions requiring sophistication and transmutation. Red may also float on the surface or come too fast and other such cautionary descriptions of the work going wrong. Although each of the main hues may be interpreted symbolically, more importantly they are aesthetically qualified, *perceived in imagistic context,* as colors in a painting derive their values from their relations with other colors. Josef Albers says:

> We are able to hear a single tone.
> But we almost never (that is, without special devices)
> see a single color
> unconnected and unrelated to other colors.
> Colors present themselves in continuous flux,

constantly related to
changing neighbors and changing conditions.[45]

The spectrum, a pantheon: "Never do the Gods appear alone," said Schiller. Polychrome, not monotone. (The Greeks *painted* their marble statues. In reverent awe we walk among their white remains, pleased with the purity; the originals would have been too gaudy.) What Albers calls the "continuous flux" of colors "constantly related" also suits the alchemical colors as a field of poetic metaphors for the transformations of soul and of nature which, as ensouled, displays its vivacity and intelligibility by means of colors.

Jung's turn to alchemy (publicly presented at Eranos Lectures in 1935 and 1936 when Jung had turned sixty) was a return to a colored soul and a colored world. It was also a return to that bifurcation in psychological history exemplified in the seventeenth century when white assumed its supremacy in science, geography, anthropology and moral philosophy, and when colonialism, black slavery, the Enlightenment and Protestant monotheism were emerging as dominants in northern-western consciousness. For only in the seventeenth century did we become truly "white." ("After about 1680, taking the [American] colonies as a whole, a new term appeared—*white.*")[46] Jung's move into alchemy re-opens these societal questions, offering both penetration to their archetypal depths and the possibility of their transmutation.

I find it curious that Jung attributes the beginnings of his alchemical research to one of his Italian dreams[47] in which a figure says to the dream ego, "Now we are caught in the seventeenth century."[48] That century saw a particularly Christian and eschatological reading[49] of alchemy—a white reading—and Jung's *Psychology and Alchemy* follows suit. So, too, Edinger.[50] But we may understand alchemy not only as a moralized recipe for a redemptive process after the Christian fashion. We may turn to Jung's alchemical work for re-immersion in history, perhaps to redirect its course.[51] For it is in alchemy that we find a self-correction of those interlocking attitudes I have been belaboring in these notes—repression and projection, oppositional thinking or splitting, superiority, invulnerability, the self-blinding of idealization, praise of ignorance, heaven-sent surety—attitudes that are archetypally given with the supremacy of whiteness.

Alchemy provides a very different vision of the supremacy of white. Because its work is called an *opus contra naturam,* it sees (and hears) through what comes naturally, first, and easily to mind. Heavenly white, innocent white and anima calmness are all too simple. The *opus contra naturam* sophisticates its own fundamentals—its elements, its metals and heat. (Not ordinary water or common mercury or direct fire—always metaphorical, imagined, reflexive, complicated.) Nothing naturalistically; everything psychologically. Hence the second white of the albedo which sophisticates the first and requires yet further transmutation.

Moreover, as Jung's great work shows, the opus aims to resolve the problem of opposites—that white devil we have been hounding throughout these notes. So, alchemy is a work against whiteness even as it strives to attain whiteness. Its way of resolving oppositional thinking is *not* by a balancing admixture of both, *not* by a golden mean between them and *not* by a transcendent third beyond them; but by desubstantiating the principle of opposition itself: the problem is solved because it does not arise. If inherent to white is supremacy and if supremacy maintains itself by denying shadow, then it is "only natural" to white consciousness to think and feel in opposites, to take them as ontologically fundamental, that is, literally. The ontological fantasy of reality as consisting of paired opposites is itself a manifestation of the white mind, and our affinity for this fantasy that mind's narcissistic defense against the wounds of self-awareness. Give up the opposites, and you can move beyond white supremacy.

In claiming archetypal priority for color, I am following not only Jung and the alchemists but also Victor Turner. After surveying the role of colors in archaic, ancient and anthropological contexts, he hazards the hypothesis that black, white and red are not only linguistically the basic color triad (cf. Berlin and Kay) or only basic to Africa (Zahan), but more: "Among the earliest symbols produced by man are the three colors" (Turner, p. 88). "They also provide a kind of primordial classification of reality." "Since the experiences the three colors represent are common to all mankind, we do not have to invoke diffusion to explain their wide distribution" (p. 90). "The point I am trying to make here is that the three colors white-red-black for the simpler societies are not merely differences in the visual perception of parts of the spectrum: they are abridgements or condensations of whole realms of psychobiological experience. . ." (p. 91). "By representing these 'forces' or 'strands of life' by color symbols in a ritual context, men may have felt that they could domesticate or control these forces for social ends, but the forces and the symbols for them are biologically, psychologically and logically prior to social classifications by moieties, clans, sex totems, and all the rest'' (p. 90).

Remarkable statements! The force of colors prior to social orders, prior to sex totems! The alchemical transmutation of the colors is one such ritual, a ritual that continues within psychoanalysis when conceived alchemically.[52] Since the domestication (sophistication) of the colors is a transmutation of the very strands of life, life's psychobiological dominants, alchemical psychology offers a mode of sophisticating the white affliction I conceive to be archetypally endemic in our civilization.

These notes remain merely diagnostic of the affliction rather than remedial unless we recognize that white, black and red as archetypal forces determining the mind can nonetheless be revisioned by it. A white mind can be inside a black skin, a red skin, any biological skin. And though we can't get out of our skins, we can go out of our minds. Alchemy describes this breakdown of the white mind as the nigredo prior to and inherent within another white.

That second white, the albedo, results from the nigredo; the only candid now psychologically tainted and shaded. This white remembers its forebears, its drop of black blood, blue blood. Its whiteness is the complete presence of all hues, not their absence, including hues yet to be born. It is a complexity of values and tints, depending on components such as lead, zinc, tin, antimony and silver, on relations and locations and temperatures, demanding the most exquisite discernment of any eye desiring to see through the blaze of first sight. As the albedo (from *albus,* ''dead white, not shining'') is inherently many-hued, so it is accompanied by the peacock's tail with many eyes (polyophthalmic *cauda pavonis*). The second white cannot be perceived with single vision, that white-eyed mind which had to be broken to bits in the nigredo. The psychic supremacy of the albedo is inseparable from its psychopathology.

''Supreme'' here takes on a psychological sense—much like daylight invisible in itself, much like echoing or mirroring empty in itself and altogether dependent upon the other for its reflection, that peculiar, secondary and inferior dependency required for insight—a supreme reflection above all others: that mirroring of itself as merely white not red, *luna* and not *sol,* silver and not gold, phase and not end; a mortally wounding reflection of the albedo's lesser rank, that it is a pale rider, a bodiless wraith, merely reflection and not the thick liquid pool, no matter how lovely, how lovable, an immaterial witness; and this supremacy of the unachieved, the weakened, the dependent is psychology's own inferior sort of supremacy sometimes called "psychic consciousness."

Psychic consciousness has moved from the seventeenth century. Yet we have been reading alchemy as if we were still in Jung's dream, as if alchemy referred only to our private eschatology, the salvation of soul, the individuation process. Alchemy—just another Christian denomination. But alchemy's principal lesson is the materiality of the psyche and that the *opus* cannot be achieved alone within the individual person. Always it is the material that transmutes its colors, the matter that sophisticates; and my subjectivity cannot be distinguished from what's out there. Alchemical psychology draws the personal soul toward the soul of things, *their* interiority and traits, *their* likes and dislikes, what they are doing and how things are going. *Anima Mundi.* I find myself in their reflection, my luck or accidents with things, the weather, what the day is like. This is the whitening of the daily, *leuke 'eméra,* good day, white day, yielding to its supremacy, the psyche all around and myself colored by its conditions. The alchemical opus takes place *in vivo* as well as *in vitro.* There is the vessel of the world which too is psyche. Which too has eyes.

That I am being watched, and watched over, we feel most vividly at night when the dayworld sleeps. I do not mean only dreams and the soul's encounter there with the intentions of the others, nor do I mean only the child's night terrors and night companions. I refer to the African experience described by Buxton: "For the Mandari night and darkness have strong emotional overtones. . . . darkness is not disturbing simply because something may be lurking in the dark, but because the darkness itself is felt to be lurking—it 'looks' at us."[53] And the Navaho chant: "I am ashamed before earth; / I am ashamed before heavens; / I am ashamed before dawn. . . . these things are always looking at me. / I am never out of their sight."[54]

This yielding to the image-pregnant materiality of the world is how I would today define "psychic consciousness." The world does not need the missionary; it is already converted, enlightened with its own opalescence. It is no longer Aristotelian or Cartesian or Christian dark matter given as blind data to the forming power of creative conscious intentions whether as the stuff for experimental physics, as slaves for colonials, as unconscious material to be made conscious. Today we will say psychic consciousness is not creative but created, and we the world's creatures. All we need do is open our eyes to its eyes.

Coda con brio
(Tailpiece without Muffler)

The seventeenth century gone and the twentieth departs. Styles of reflection reflect styles of centuries. Though Jung's formula, reflection equals consciousness (*CW* 11, §235 n.9; *CW* 8, §§241–43) may still hold, we have come to imagine reflection differently. "Post-modern" they call it. Leaving the twentieth means leaving the modern mode of reflection defined as bending back and away from the object toward interior subjectivity where images are formed that 'refer' in some way or other to the object. Yet to bend back and away in favor of the mind's deliberations and interpretations retreats from the other, the very ground on which reflection depends. So how do we know that what's going on in reflection refers to what's out there?

Faced with the paradoxical isolation of its reflections, modern consciousness invented the referent, insisted that its observations were indeed about the other, that its reflections truly reflected something literal, opaque, outside itself—the referent. But modernism could not escape the white of its own eyes: to be conscious was to be separated from and utterly untainted by anything but itself. Here lay the paradox requiring the scientific method, ever refining its objectivity to reach beyond the solipsism intrinsic to its style of reflection. For whiteness is supremely solipsistic, like the perfect child or Aristotle's God needing no friend, like the calm, smooth anima, self-satisfied, its delusions self-consistent, resting in a unified field on a coherence theory of truth. Extraversion, materialism, utilitarianism, positivism, projection itself—each shows the obsessive-compulsive agony of the white mind trying to get out of itself, desperately seeking the obscure object of its desire. A solipsistic history repeating the same moralistic shibboleth: Love thy neighbor; do unto others; connect, only connect; relate, relate. . . .

As modern psychology recognized this double delusion—that its reflective consciousness does not really require another and that this consciousness really does refer to another—it had to divide the mind. It had to invent the unconscious in order to remind consciousness that it could never be as white as it wished. The "discovery" of the "unconscious" came as a late stage of modernism, indicating its decline by turning its projective roots back onto itself. The "discovery" was actually a self-discovery, a backhanded welcoming of reflective consciousness's own delusional base, turning the delusion into irony and joke, a way to look back over its own shoulder, to reflect its own downfall, to become "post." It forced white consciousness out of its supremacy, saying to it: "You can't know what you are talking about," asking, "What are you referring to?" And though post-modernism loves to deny the referent, this move does not deny what's out there. It does deny that consciousness is the way to it. Modern consciousness as white consciousness cannot get there, and what it finds is always dead. Having walled itself in, it blames it on the wall. (And how it listens for signals, pining for life somewhere in space!)

To deny the referent is merely to state that what's out there is not knowable as referent or ever to be referred to as referent—tied to my consciousness for its significance, as if on its own each thing were a dumb wall, irredeemably unconscious, signifying nothing. Rather, to deny the referent affirms the thing. What's there is a first class act *an sich,* playing dead to elude projections; a sparky, flickering thing with its own rights, proving its mercurial freedom by escaping every frame of reference into another and another, sheltered by its friends, other things, among which it lives its life and into whose company I may be allowed as a thing among things asleep by the wall.

Formerly, reflection meant placing an event against a background—systematic schema, symbolic meaning, mythical topos. Post-modern reflection displaces, shatters the confident prism, meaning untied to references, dislodged from backgrounds and structures, allowing new falls, free falls. Falling from the heaven of white certainty, out of the arms of the opposites. Through the looking glass and into the images. "Raising consciousness" now requires an *abaissement du niveau mental;* Lucifer, the has-been star, now guides the epistrophé because the way down is the way up. The knower becomes known through the statements of his knowing. Now it's the overseer in the woodpile, *is* the woodpile. Not projection, subjection. They've got their eyes on me; I am their referent, their text. "I" have fallen out of my mind, out of the twentieth century, no place to hide and everywhere to go. A goalless going, without a project, 'unconscious,' so that we can never ascertain arrival. The fall may be continuous but not abysmal, not a bottomless, horrific maw, and the traveling not labyrinthine; instead, a lovely continuing along and there is always that wall with its faces. Such could be consciousness in a post-modern psychic style. Post-modern optics—the study of scotoma, of blindness, of the ways in which the eye that sees cannot see itself. Epistemology—the study of how I am shown to and known by the sleepless Argus-eyes reminding my mind of their minds, and their constant minding.

White consciousness already feels itself fallen, feels itself shadowed, under surveillance, poisoned, exploited, assaulted and abused. Return of the slave, return to the native. A victim consciousness arises as the century departs. In keeping with the old modernism we interpret identification with the victim personalistically, try to cure it, to regain supremacy. But there is another reading: white consciousness forced into the postmodern psyche, into an animistic recognition of shades everywhere even in the brightest day and closest love. Melanoma under the white skin. In white itself there lurks invisible disease, uncontrollable terror, impersonal rape and spies who have infiltrated our closed subjectivity. We cannot erect the impenetrable defensive shield, the white immune system no longer reliable. Modern consciousness now suffers its own uncertainty principle, pressed to recognize that all its reflections do not extend beyond the prism, the prison, of its own unconsciousness, no matter how expanding the universe, how luring the moon.

No other place now for the unconscious except inside the act of reflection so that psychic consciousness in this time of the fall can be defined only ironically as self-reflexive, where reflection refers to the many-eyed ways each act of consciousness is supremely multiple and thereby supremely self-limiting, inevitably, humiliatingly projective and indestructibly destructive, or (let us say) supremely unconscious.

1. Brent Berlin and Paul Kay, *Basic Color Terms: Their Universality and their Evolution* (Berkeley and Los Angeles: University of California Press, 1969).

2. Counter examples are listed by D.L. Snow, "Samoan Color Terminology," *Anthropological Linguistics* 13/8 (1971): 389. Cf. the following articles in the same periodical: R.W. Wescott, "Bini Color Terms," 12/9 (1970): 349–60; J.A. Frisch, "Mohawk Color Terms," 14/8 (1972): 306–10; H.B. Broch, "A Note on the Hare Indian Color Terms," 16/5 (1974): 192–96.

3. Eleanor Irwin, *Colour Terms in Greek Poetry* (Toronto: Hakkert, 1974); Liddell and Scott, *A Greek-English Lexicon* (Oxford: Oxford University Press).

4. C.T. Lewis and Ch. Short, *A Latin Dictionary* (Oxford, 1894).

5. Irwin, *Colour Terms,* pp. 163–65.

6. Ibid., p. 166 (cf. Plutarch, *Pericles* 27).

7. Dominique Zahan, *Religion, Spirituality and Thought of Traditional Africa* (Chicago: University of Chicago Press, 1979); "White, Red and Black: Colour Symbolism in Black Africa," in *Color Symbolism: Six Excerpts from the Eranos Yearbook 1972* (Dallas: Spring Publications, 1977), pp. 55–80.

8. Victor Turner, *The Forest of Symbols* (Ithaca: Cornell University Press, 1967; paperback 1970), pp. 69–70.

9. Ibid., pp. 76–77.

10. Geoffrey Parrinder, *West African Religion* (London: Epworth Press, 1949), p. 119.

11. Jean Buxton, *Religion and Healing in Mandari* (Oxford: Clarendon Press, 1973), p. 385.

12. V. Turner, *Forest of Symbols,* p. 84 (Cherokee).

13. Gladys A. Reichard, *Navaho Religion,* Bollingen Series (New York: Pantheon, 1950), 1 : 189.

14. V. Turner, *Forest of Symbols,* p. 82 (Madagascar).

15. Ibid., p. 83 (Malay Peninsula).

16. Heinrich Zimmer, *Philosophies of India,* Bollingen Series (New York: Pantheon, 1951), pp. 229–31&n.

17. Cf. Frederick Turner, *Beyond Geography* (New Brunswick: Rutgers University Press, 1983), p. 160.

18. Fernand Braudel, *The Perspective of the World,* 3 vols. (New York: Harper & Row, 1984), 3 : 434.

19. Ibid.

20. Against this archetypal account, structural anthropology argues that no single color like white has intrinsic meaning: a meaning is often reversible depending on context and other factors. The true bi-polarity of color into an opposite meaning is not black *vs.* white or white *vs.* red, etc., but the presence/"absence of colour or between two chromatisms of different degrees [stressed/unstressed]. These observations seem to make it possible to dispose of theories making use of the concepts of 'archetypes' or a 'collective unconscious'. It is only forms and not contents which can be common" (C. Levi-Strauss, *The Savage Mind* [Chicago: University of Chicago Press, 1966], p. 65). The opposition between content and form in Levi-Strauss's theory of mind as well as his monotheism—i.e., the one basic idea of binary opposition in which the significance of the poles may be reversed without altering the significance of the structure—raises the question whether "binary opposition" is purely formal or whether it too isn't a content, with denotative meaning intrinsic to mind everywhere, and thus itself 'archetypal.'

21. Cf. James Hersh, "From *Ethnos* to *Polis,*" *Spring 1985:* 60.

22. Cf. J. Hillman, "Silver and the White Earth (Parts I and II)" in *Spring 1980* and *1981.*

23. Cf. Irwin, *Colour Terms,* pp. 112–35, on "white women" as a Greek literary convention.

24. Ibid., p. 151.

25. Ibid., p. 185.

26. Ibid., p. 186.

27. Theophrastus, *On the Senses,* ed. G. M. Stratton (Amsterdam: Bonset, 1964), §73.

28. Cf. J. Hillman, "Alchemical Blue and the *Unio Mentalis,*" *Sulfur* 1 (1981) : 33–50.

29. M.-L. von Franz, *Aurora Consurgens* (London: Routledge, 1966), p. 243.

30. "For there are never in nature two beings which are exactly alike and in which it is not possible to find an internal difference. . ." (Leibniz, *Monadology,* 9). Cf. "Identity" in *The Encyclopedia of Philosophy* (New York and London: Macmillan, 1967), 4 : 122 a&b.

31. F. Nietzsche, *The Dawn of Day,* §460.

32. K. W. Schaie, "The Color Pyramid Test," *Psychol. Bull.* 60/6 (1963) : 540–41.

33. J. Buxton, *Religion and Healing in Mandari,* p. 389.

34. F. Braudel, *Perspective of the World,* 3 : 440, 435, raises questions about African slavery that existed in monstrous proportions ("900,000 in the sixteenth century, 3,750,000 in the seventeenth, between 7 and 8 million in the eighteenth" sent off to the New World alone). Yet "Africa had already developed this bad habit long before the Europeans arrived, sending slaves to Islam, the Mediterranean and the Indian Ocean. Slavery was endemic to Africa, part of the structure of everyday life. . . ." Could we substitute the sociological word "endemic" for the psychological word "archetypal"? Was the terrible possibility of slavery given, not only with the social and economic structures in the tribal villages, but in the archetypally "inferior" meanings of black employed by the Africans themselves?

35. Calvin Harlan, *Vision and Invention* (Englewood, New Jersey: Prentice Hall, 1986), p. 84:

"... *shades* (mixtures of greater or lesser quantities of colors with black)."

36. Winthrop D. Jordan, *White over Black: American Attitudes toward the Negro, 1550–1812* (New York: Norton, 1977), p. 4.

37. Braudel, see n. 34 above.

38. Jordan, *White over Black,* p. 7.

39. For a detailed survey in parody of white folk attitudes, see Martin Mull and Allen Rucker, *The History of White People in America* (New York: Putnam/Perigree, 1985).

40. F. Turner, *Beyond Geography,* in the chapter "Possession" presents examples of whites who chose the other side and the fury released by this "betrayal" of ethnic identity.

41. Cf. Gilbert Durand, "Psyche's View," *Spring 1981:* 2–5 on seventeenth-century "Apollonianism."

42. *The Freud/Jung Letters,* ed. Wm. McGuire (Princeton: Princeton University Press, 1974), 255F, 12 May 1911.

43. See, for instance, Henri F. Ellenberger's massive classic, *The Discovery of the Unconscious* (New York: Basic Books, 1970).

44. Cf. Theodore Roethke, "In a Dark Time," *Collected Poems* (1966).

45. Josef Albers, *Interaction of Color* (New Haven: Yale University Press, 1963; 1975 edition), p. 5.

46. Jordan, *White over Black,* p. 95.

47. On Jung's Italian dreams and "Italian complex" see my *Loose Ends* (Dallas: Spring Publications, 1975), pp. 160–61 with notes.

48. C. G. Jung/Aniela Jaffé, *Memories, Dreams, Reflections* (London: Collins and Routledge, 1963), p. 195.

49. Cf. Stanton J. Linden, "Alchemy and Eschatology in Seventeenth Century Poetry," *Ambix* 31/3 (1984): 102–24.

50. Edward F. Edinger, *Anatomy of the Psyche: Alchemical Symbolism in Psychotherapy* (La Salle, Ill.: Open Court, 1986).

51. The notion that history is open and may be redirected is essential to the psychological approach since all 'past facts' are always subject to psychological re-view. Reconstruction, interpretation, biographical integration—those essentials of every psychoanalysis depend upon the notion that *history,* as Henry Corbin insisted, *is in us, not we in history.* Corbin's reversal of the usual Western view of *history* as a collection of implacable outside forces and facts determining human life compares with Jung's reversal of the usual Western view of *psyche* as a private subjectivity inside the human skin. Instead, says Jung, *not the psyche in us; we are in the psyche.* Together Corbin and Jung allow us to turn psycho-history inside out. Psyche embraces history; history is an activity taking place in the psyche. As the individual works on the soul, so he re-immerses himself in history redirecting its course, a process we can witness in the working through of any case history in a prolonged analysis. Corbin and Jung together point the way beyond Nietzsche's eternal return, for only that history returns which has not been put through the alchemical operations and relocated in the psyche. Only then, as Santayana (and Freud) said, are we forced to repeat because history has become literalized. Only then it still stands outside and thus stands still as traumatic fact.

52. "... Chaldeans used to make images which had an 'antipathetic' effect upon the 'streams of Heimarmene' [destiny]; that is the demons subservient to Hecate, who were considered as agents of diseases. The Chaldeans employed for this purpose a three-coloured earth.... the statuettes of Hecate made by the magicians were of white, red and black wax.... Hecate the mistress of the three elements: the fiery ether, the white radiant air and the dark earth.... It seems probable that the three colours of the statues made by the Chaldeans were meant likewise to represent the

elements of the world (and perhaps the three Chaldean world-circles: the empyrean, the ethereal and the hylic)'' (Hans Lewy, *Chaldean Oracles and Theurgy* [Paris: Études Augustieniennes, 1978], p. 292).

53. Buxton, *Religion and Healing in Mandari,* p. 387.

54. Quoted in full in P. Berry, *Echo's Subtle Body* (Dallas: Spring Publications, 1982), p. 5.

THE SHADOWS OF THE HOLISTIC EARTH

PETER BISHOP
(Magill, South Australia)

The 1960s saw the first photographs of the Earth as a planet, and in 1968 it was revealed to its inhabitants by a direct television broadcast from an orbiting spacecraft. Concerning this, Metzner writes:

> The first photographs of the whole earth returned from space signaled the beginning of a new cycle of all-inclusiveness: there she hung like a blue-green jewel in the velvet black of deep-space, laced with sparkling atmospheric veils—our spaceship, our mother, our planet. The world is one. We are all together now.[1]

The abrupt creation of this idealized image of wholeness, the "Whole-Earth" fantasy, is an event unique to the industrial cultures. It is a complex holistic image of the Earth: physical, cultural, spiritual, its history and its future. "Space-ship Earth," "Earth Inc.," "Global Village," "the Earth as a cell," "Whole-Earth," "the Mandalic Earth"—all portray the Earth as a discrete evolutionary unit in an immense cosmic drama. But the confidence, exuberance and sense of rightness inspired by this imagistic event have obscured imaginative paradox. By what routes did the Whole-Earth image acquire its present massive coherence? At what moments within the emergence of this image have confidence and hope coincided with anxiety and despair?

Evolution and the Melancholy of Immensity

With the publication of Charles Lyell's *Principles of Geology* in 1830, Western humanity was confronted by a past that spanned millions of years rather than a mere six thousand. This great void cried out to be filled by

---

The author is lecturer in Sociology at the South Australian College of Advanced Education. Other articles by the author—*Archetypal Topography* and *The Imagination of Place: Tibet*—have appeared in *Spring 1981* and *1984*.

the imagination. As Hitchcock wrote in his 1849 essay "The Poetry of Geology," "Shall not geology, which is the first science in affording scope for the imagination . . . afford themes for the poet?"[2]

Such scale and duration, unpopulated by traditional imaginings, posed a grave threat to human vanity and identity. Even the fantasy landscapes that rose to fill these vast gulfs of time had a sense of gloom and melancholy about them, populated as they were by ponderous dinosaurs and brutish ancestors, and covered by dark forests and teeming swamps.[3] The history of evolutionary theory is an archeology of imaginative landscapes, a chronicle of enthusiastic debates among contending landscape fantasies: "If we will imagine still further, let us call up the shades of the *mighty dead,* the Iguanodons, Megatheria, and Dinornithes, and hear them tell their story concerning their former habitations."[4]

From the mid-nineteenth century a proliferation of "Lost Worlds" adventure tales attempted to integrate and diffuse the monotonous threat of these alien regions.[5] Heroic struggles against primeval dangers provided a *human* focus in planetary landscapes that otherwise threatened to reduce humanity to insignificance. Such fears were a quiet but constant companion to the optimism and confidence in early evolutionary texts.[6]

Lyell's work also unlocked a parade of immense *future* landscapes. By the 1930s optimistic planetary evolutionism had waned. Both Teilhard de Chardin by a kind of mysticism and Olaf Stapledon in fiction consciously expressed the dark ambivalence of such fantasies. Stapledon, that seminal mythologist of immensity, wrote,

> We desire our present civilization to advance steadily toward some kind of Utopia. The thought that it may decay and collapse . . . is repugnant to us. Yet this must be faced as at least a possibility. And this kind of tragedy, the tragedy of a race, must . . . be admitted into any adequate myth.[7]

Stapledon's visions are saturated with the melancholy of immensity:

> A strange passion of loneliness seized me. Like a prisoner I could have battered on the walls of the cosmos, had there indeed been any walls. But I was imprisoned in a boundless finitude.[8]

Even Teilhard de Chardin's otherwise enthusiastic imaginings could not avoid such disturbing reflections: "The whole psychology of modern disquiet is linked with the sudden confrontation with space-time." This "malady," he wrote,

> manifests itself as a rule by a feeling of futility, of being crushed by the enor-

> mities of the cosmos. . . . Enormity of duration—sometimes having the effect of an abyss . . . and at other times . . . the despairing effect of stability and monotony.[9]

It is surely no coincidence that with the advent of vast geological time and an even more immense cosmic arena there arose a desire for a harmonious planetary image. A struggle ensued to imagine a stable planetary sphere evolving in a steady and well-organized process. Through the nineteenth century, study shifted from a six-thousand-year period of planetary catastrophes to the immensity of a steady and unchanging process. Catastrophism as a geological theory was ridiculed, and the stability and coherence of planetary evolution became idealized in the face of immensity. Melancholy and endurance went hand in hand.

Buckminster Fuller wrote of the Earth's television image as filmed by an orbiting spacecraft:

> As the planet Earth's cloud-bewreathed sapphire-watered blueness was dramatically displayed against the x-trillions of time years of nothingness . . . [the] commentator exclaimed, "There she is—floating there!"[10]

Such calm oceanic metaphors are common, but paradoxically the sudden visual coherence of the Earth's gestalt has coincided with a resurgence of catastrophism as an explanation of the formation of planetary life. Such ideas are now openly expressed on the fringes of evolutionary speculation. One popular author has even suggested that the Earth can easily be turned upside down by the influence of only a small interstellar body and that this has already occurred several times within recorded history.[11] Yet whether tumbling, floating, swimming or just hanging in space, the exemplary image of the globe is not disturbed.

Bachelard begins a chapter on "Intimate Immensity" with the quotation "Space has always reduced me to silence."[12] He continues, "If we could analyze impressions and images of immensity . . . we should soon enter into a region of the purest sort of phenomenology—a phenomenology without phenomena. . . ." Only in a state of purest reverie can our imaginations enter such regions without recoiling. The "Holistic Earth" relieves the burden of a formless immensity. This stable global image gives focus to the boundlessness of space. Futility, monotony and melancholy are the consistent results of such an extended imagining.

Ecology and the Dread of Insects

Once the planet as a *system* rather than the evolutionary *process* becomes the prime focus, then we have shifted into the fantasy of ecology. Different hopes and fears arise, new coherences are mobilized. The word "ecology" first appeared in the English language in 1873.[13] Ecology lays emphasis on the whole rather than the parts, on the interaction between organism and environment. Harmony, balance, interdependence, unity, totality are used time and again to describe idealized systems. Life becomes coherence, the Earth a global cell. Humanity is imagined as merely one life form among many, a planetary-being inseparably enmeshed in a living web.[14]

Whilst the field of ecology was initially shaped by nineteenth-century biology and botany, in the twentieth century it has grown to encompass all disciplines from the arts to the social and physical sciences. "The only thinking worthy of the name must now be ecological," writes Mumford.[15] The organic and inspirational metaphors of those who proclaim a holy global unity—"Integrity is wholeness, the greatest beauty is organic wholeness. . ."—contrast with the systems language of the holistic pragmatists: "God . . . [is] an abstract love-momentumed gyrocompass. . . ."[16] Seemingly opposed to both are the images of an *unholy* unity. For Karl Marx the globe was indeed becoming united, but by capitalism and imperialism. Similarly, Margaret Mead writes that humanity has itself "woven the previously dispersed and unconnected population of the planet into a single, interconnected, mutually dependent and totally-at-hazard single group."[17] The global village euphoria of the 1960s seems naive alongside the contemporary shadow image of a "totally-at-hazard single group."

In his recent popular study *The Fate of the Earth*, Jonathan Schell comments gloomily that after an atomic holocaust all that may be left is "a republic of insects and grass."[18] This is a modern restatement of an ancient fear, that the insect kingdom will displace Homo sapiens as the dominant species and inherit the Earth. While the survival of birds or dolphins might offer slight consolation, the knowledge that planetary dominion would probably pass to the insects, especially ants and cockroaches, evokes disgust, despair and desolation. Insects have long been associated with the devil. In the Jungian tradition, a crowd of insects in a dream is often taken to symbolize a latent psychosis, a fragmentation of the basic personality. James Hillman suggests that they evoke the rejected autonomy of the Western psyche; insects in dreams present the natural mind or intelligence in the complexes.[19]

Holism fears fragmentation. It uses insects to evoke aggressive fantasies. Homicide is a crime, insecticide a household product. Insects inspire a panicked shooting from the hip with toxic aerosols. Hillman comments: "Often, when the bug appears, so does fire." Insect fantasies touch upon a fear of disintegration, of pollution, of a loss of identity. They express fears of both an omnipotent, well-organized system and of the chaotic, faceless multitude. The dark images specific to alienation and the industrial age—the masses, bureaucracy, overpopulation, totalitarianism—have all at some point been expressed in entomological metaphors: hives of industry, ant-like city commuters. Even the information-systems language of modern biology and cybernetics has the feel of insects and their organization: units, bits, microchips, and of course "bugs." To imagine the Earth as a single cell, as a holistic unity, summons dreams of insects.

Yet insects are also mythologically associated with sorting out, sifting through, an attention to details. They are messengers of the underworld insisting that we move from our glorious abstractions of global unity into a fragmented imagining that is more Earth-bound. But to the onesided imagination of holism the insects are nibbling away everywhere, in the foundations of everything. Fantasies of global unity have been increasingly undermined by a pervasive sense of crisis. Everything is breaking down, everything seems to threaten us: the air we breathe, the water we drink, the food we eat. There is a profoundly diversified anxiety. The *reverse* side of global unity reveals the face of fragmentation, panic and crisis—not a single big one in the future but innumerable little ones, now, everywhere and every day. Crisis reporting, crisis intervention, crisis management, crisis care, crisis counseling are all peculiar to life in late twentieth-century Western society.[20]

Concern about ecology seems to promote ceaseless activity. The problems are always presented as *urgent,* the question always what to *do, now!* Contemplation has been replaced by activism. As one ecological activist exclaimed: "We used to be studying guillemots: now we are trying to save the human species."[21]

An 1898 lecture, "Man's Dependence on the Earth," resignedly proclaims, "Man can never burst the bonds that subject him to nature."[22] Loren Eiseley wrote that "the discovery of the interlinked and evolving web of life" was pre-eminent among the causes of "an entirely new and less tangible terror."[23] And Jung warns, "It could easily be conjectured that the earth is growing too small for us, that humanity would like to escape from its prison. . . ."[24] Humanity seems trapped within its own dreams of unity. Insect fears of chaos, fragmentation, and loss of identity are complemented by those of being trapped, suffocated, imprisoned. The *web* of life, the eco-*net*work, summons up such fears.

The widespread celebration, even worship, of humanity's inescapable participation in the web of life belies the death and destruction threatened by the spider at the center. The web is an appropriate symbol for the shadow side of the much proclaimed "return to Mother Earth." The web is not only a holistic image to be contemplated in wonder but a labyrinth down which humanity stumbles after a sense of its own identity and security.

An Immense Loss

Planetary holism does not simply present an exemplary image of order but also of immense loss. The sheer size and absoluteness of contemporary fantasies of loss are unprecedented:

> There is no survivor, there is no future, there is no life to be recreated in this form again. We are looking upon the uttermost finality which can be written, glimpsing the darkness which will not know another ray of light. We are in touch with the reality of extinction.[25]

Species have become extinct throughout the Earth's history, and the experience of extinction will have been consciously faced by many peoples. But no species has ever been forced to contemplate the distinct possibility of its own extinction. "Extinction," "annihilation" crush imagination. It is difficult to view as fantasy the prospect of the immediate extinction of the human species, to see it psychologically.

In 1848, just four years after the last two living Great Auks were clubbed to death and their solitary egg crushed by the boot of an Icelandic fisherman, Karl Marx wrote of humanity's "species-being." It was then a difficult concept to grasp. Three years later the Crystal Palace exhibition of 1851 was dedicated by Prince Albert with the words:

> Nobody who has paid attention to the peculiar features of the present era will doubt for a moment that we are living at a period of most wonderful transition, which tends rapidly to accomplish that great end, to which, indeed, all history points—the realization of the unity of mankind.[26]

One hundred and thirty years later we realize that global unity has its shadow side in global war and global pollution. Yet emerging from beneath this dark underside is the awareness of ourselves as a particular species. Today it is much easier to grasp Marx's image of a species-being. To imagine our identity in this way means facing death. To imagine species-existence necessitates the imagining of species-extinction; they are two sides of the same image. To the average Westerner of Marx's day, both the extinction of the peaceful Great Auk and Marx's image of species-being would have seemed alien, abstract and remote. To the contemporary ear they have a modern ring.

Imperialist expansion in the nineteenth century brought with it a parade of colorful new images—races, ruins, costumes, languages, religions, customs, plants, fossils, animals, insects and birds. Botanical gardens, zoos, illustrated weekly magazines, magic-lantern shows and colonial exhibitions all sprang up and brought these images back to the centers of the empires. Returning explorers were feted, given royal receptions and showered with scholarly honors. Institutions such as the Royal Geographical Society were set up to promote exploration and control the flood of images which threatened to rend the intellectual, imaginative and moral fabric of the era. On the backs of these countless new images came an awareness of the *immensity* of death. The discovery of the ruins of numerous, unimagined ancient civilizations from the Himalayas to the Gobi Desert, from Crete to Central America, from Arabia to Africa underlined the transience of the nineteenth-century European empires and civilizations. The beginning of archeology and anthropology was also the beginning of a mourning for our own culture, a lament for the cities, languages, customs and daily life of Western civilization.

Planetary biology and botany brought with them not just amazement at the apparently endless parade of species but an awareness of the vast continuity of species-extinction. For each new species discovered, dozens had vanished, and as Homo sapiens's imagining of itself became more immersed in material nature's web, a common fate has increasingly been its shadow. The past forty years have seen both an acceleration in the imagined scale of loss and a fearful collapse of the span of time available. The future eons confidently created for humanity by evolutionary fantasy have been radically curtailed. The imagination of time has now undergone two revolutions in less than 200 years. The one that began with Lyell has been overtaken by the one that began with Hiroshima. Alongside the vast past and future timescape of evolutionary imagining but absolutely at a remove from it is the continual present of nuclear extinction. It now seems that

Homo sapiens has either an immense future or none at all. Such an image creates a swing from cosmic ennui to planetary panic.

Not only human memory but the very ground of imagining is now threatened with extinction. Each of the imagined levels of loss—individuality, civilization, human species, the animal and plant world, the matrix which creates and supports life in the form we know it—expresses fears about the absence of psychological referencing, the death of *memoria.* Through these terrible images of loss, we are pulled down into the soul of things, the *Anima Mundi.*[27]

Wilderness and the Loss of Beauty

Wildernesses have been called "meccas for a pilgrimage into our species past," "reservoirs of human freedom," "part of the geography of hope."[28] Wilderness preservation has been urged for many reasons—as a rejection of industrial civilization and a return to primitivism, as repositories of invaluable scientific data, as training grounds for the development of self-reliance and survival skills, as places that enhance team spirit, as solitary retreats for contemplation and worship, as centers of healing where the stress and confusion of urban life can be released, as salutary correctives to anthropocentrism. Yet the basic property of wilderness is aesthetic.[29] A fundamental anxiety about beauty lies at the heart of wilderness imagining.

While advocates of wilderness consistently extol its beauties, contemporary appreciation scarcely matches the delight and awe it evoked in earlier days. Rousseau complained that the voluptuousness of nature overwhelmed him; Thoreau confided, "My senses get no rest"; Muir insisted that beauty was as important as bread. On his first walk among the Himalayas, Younghusband ecstatically exclaimed, "Oh Yes! Oh Yes! This really is splendid! How Splendid! How Splendid!" For us such places are now always touched with impermanence; they are reminders of loss.[30] But the imagination of wilderness still gives the image of a holistic Earth a sensual cohesion. It provides a preeminently *visual* experience of planetary life. Through the eros of beauty, both the ecological and evolutionary visions are restrained from excessive abstraction.

As seen from space, the image of the Earth elicits a scarcely veiled ache: "Wrapped in a mantle of cloud, she swam in loveliness through the sea of space. It is our choice to violate or to grace the beauty that is hers."[31]

While ecology is concerned with the breakdown of connections and the destruction of an ancient and fundamental harmony, wilderness conservation consistently uses metaphors of violation, rape and despoliation.

But perhaps the great question of our time is less the alienation from "nature" than from "beauty." "Wilderness" originally had connotations of disorder and bewilderment. There is a madness about wilderness. The "savage and dreary" wilderness of Maine shocked Thoreau and forced him to temper his previously onesided enthusiasms. Few would share his reaction today. Are we more enlightened in proclaiming that such places no longer disturb us, or have we lost touch with some fundamental paradox? Wilderness activists frequently insist that their aim is to reeducate the West into an unequivocal appreciation of the beauty of "natural" wilderness. Such a goal is puzzling, for it is unlikely that any human culture has unparadoxically enbraced *all* its known regions of landscape.[32] There is a concerted effort to clean up the imagination of wilderness and to remove any gloom associated with it. In the past this was achieved by literally clearing away the forests, swamps and jungles or by cultivating the deserts. Now it is being fantasized away by insisting that the oppressiveness and fear traditionally inspired by wilderness were biased and wrong.[33]

The very idea of wilderness had its origins in an oppositional fantasy: wild/tamed, savage/civilized, ungodly/godly, madness/sanity, chaos/order, confusion/harmony.[34] The image of wilderness has always been used as something to define and identify oneself by. It always invokes *another place* and can only exist by being set apart from somewhere else. Wildernesses mark the frontiers of an imagining of extremes. As unambiguous shrines of natural beauty, they are also contemporary entrances to the underworld, to landscapes where such beauty is imagined to be totally absent.

There is a danger that onesided fantasies will confine "natural beauty" to so-called wilderness areas and exclude it from everyday life in the city. Already wildernesses are being packaged for the quick high. They are quickly becoming a closed text: e.g., environmental psychology has already defined and proclaimed *the* "wilderness experience," which has become the focus of attention of educators, therapists, social workers and so on.[35] Ways have been devised to heighten its intensity and channel its direction. Presumably Thoreau would need reeducating in his attitude towards the wilderness landscape of Maine. The "wilderness experience" is taking shape as yet another programmed, instant consumer item. The wildernesses of the world are becoming either the adventure playgrounds of the wealthy nations or the cathedrals of a new dogma.

We bypass paradox either by chopping down ancient trees or uprooting "undesirable" Western fantasies. As the wilderness becomes a onesided place of salvation, fantasies of evil shift over to the city, the machine, the mining company. Already the variety of possible imaginative reactions to the wilderness is being constricted into the smug morality of holism and the packaged therapeutic milieu of the twentieth century.

Victim of the Shadows?

The routes by which the holistic planetary image arrived in Western consciousness were consequently not free from shadow. From evolutionary imagining comes images of humanity crushed beneath the burden of a remorseless continuity or negated into a state of despairing ennui by a limitless expanse of time and space. The ecological imagination embraces fears of fragmentation, chaos, of imprisonment within the web of life, with a loss of human identity. The imagination of wilderness is tinged with nostalgia and with the mood swings that always come with heavily literalized and unreflected fantasies about beauty. These are some of the deep pathologies specific to our time, though they commonly appear as dreams of insects or as irritating concerns about the minutiae of our everyday lives.

The well-known anthropologist Mary Douglas concludes her study of contemporary environmentalism by raising the question "why America is more passionately involved than any other Western nation in the debate about risks to nature."[36] Does a position at the center of global power allow a privileged insight into global issues, or does such an on-top position encourage and necessitate global imaginings? When power and manipulation are global, so are the images of hope and meaning. Global scenarios, global solutions, global problems are all part of the same specific fantasy.[37]

Eiseley insists that humanity "must in fact have walked the knife edge of extinction for untold years. As he defined his world he also fell victim to the shadows that lay behind it."[38] It seems the nightmare will be whispering at our shoulders for quite a while. The important question is not one of accidental mechanical error, a technical fault, a malfunctioning microchip in an early-warning installation. The basic issue is whether we can bear the unrelieved intensity of our own nightmares. As these become less easily dismissed from our daily life, they congeal into an image of darkness that is just as exemplary, dense and coherent as that of the holistic Earth.

A Fragmented Ecology

Bachelard wryly remarks, "[Adults] demonstrate to the child that the earth is round, that it revolves around the sun. And the poor dreaming child has to listen to all that! What a release for your reverie when you . . . go back up the side hill. . . ."[39] A specific, fragmented imagining in no way underestimates the threat to the planet nor does it reveal any less concern about the environment. Certainly the image of a holistic Earth points to the urgent need for imaginative vessels to hold, cook and digest the fantasies of our time. But fragments also heal. The questions posed by a global imagining are in themselves *shattering.* They consistently fragment the comfort one might take in a premature holism. As we have seen, there is an inherent shadow of destruction in idealistic holism. We need to descend into these shadows of the holistic Earth, for this image was born simultaneously with one of its own doom. Eiseley writes of "the dark murmur that rises from the abyss beneath us and that draws us with uncanny fascination." These murmurings are the world calling attention to itself, reestablishing itself as a psychic reality.[40]

1. R. Metzner, *Maps of Consciousness* (New York: Collier Macmillan, 1979), p. 2.

2. In R. Hazen, ed., *The Poetry of Geology* (London: George Allen and Unwin, 1982), p. 14. For discussion about the fluctuating scale of immensity, see S. Kern, *The Culture of Time and Space* (Cambridge: Harvard University Press, 1983), pp. 37–38.

3. E.g., see S. Prickett, *Victorian Fantasy* (Bloomington: Indiana University Press, 1979). S. Kern cogently argues that a revolution in Western conceptions of time and space occurred between 1880 and 1918. He traces such an imaginative shift from art to imperialist geographical aggrandizement, from theoretical physics to psychoanalysis (Kern, *The Culture*).

4. Hazen, *The Poetry of Geology.*

5. E.g., A. Conan Doyle, *The Lost World* (London: Hodder & Stoughton, 1912); and E. Rice Burroughs, *The Land That Time Forgot* (New York: Story Press Corporation, 1918).

6. See the comments by Hooker to his friend Darwin in L. Huxley, *Life and Letters of Sir Joseph Dalton Hooker,* vol. 11 (London: John Murray, 1918), p. 258.

7. O. Stapledon, *Last and First Men* (1930) (London: Methuen, 1978), p. 11.

8. O. Stapledon, *Nebula Maker* (London: Sphere Books, 1979), p. 130; see also his *Star Maker* (London: Methuen, 1937).

9. P. Teilhard de Chardin, *The Phenomenon of Man* (London: Fontana, 1975), pp. 249–50, and also J. Huxley's essay in the same volume.

10. R. Buckminster Fuller, *Earth Inc.* (New York: Anchor Press, 1973), p. 128.

11. F. Hitching, *The Neck of the Giraffe: Or Where Darwin Went Wrong* (London: Pan, 1982), pp. 164–65.

12. G. Bachelard, *The Poetics of Space* (Boston: Beacon Press, 1969), pp. 183–84.

13. See B.E. Fernow, "Applied Ecology," *Science* 17 (1903); V.M. Spalding, "The Rise and Progress of Ecology," *Science* 17 (1903).

14. E.g., A. Chisholm, *Philosophers of the Earth: Conversations with Ecologists* (London: Sidgwick and Jackson, 1972).

15. In Chisholm, *Philosophers.* See also G. Sessions, "Ecophilosophy, Utopias, Education," *The Journal of Environmental Education* 15/1 (Fall 1983).

16. The quotations are from Robinson Jeffers and Buckminster Fuller (*Earth Inc.*, p. 69).

17. Quoted in M. Douglas and A. Wildavsky, *Risk and Culture: An Essay on the Selection of Technical and Environmental Dangers* (Berkeley: University of California Press, 1982), p. 64. They suggest there is little difference between the way industrial society views environmental issues and the supposedly "magical" world view of pre-literate societies. See also B. Weisberg, *Beyond Repair: The Ecology of Capitalism* (Boston: Beacon Press, 1971) on the unholy unity.

18. J. Schell, *The Fate of the Earth* (London: Picador, 1982), pp. 63–65.

19. J. Hillman in "Going Bugs" (one of his lectures on "Animals in Dreams" at The Dallas Institute of Humanities and Culture, Spring 1982) also referred to *The Lord of the Flies,* the chorus of insects in *Faust,* and Kafka's *Metamorphosis.*

20. See R. Sardello, "The Suffering Body of the City," *Spring 1983:* 145–64.

21. The ecologist Dr. N. Moore in Chisholm, *Philosophers;* see also I. Barbour, *Earth Might Be Fair: Reflections on Ethics, Religion, and Ecology* (New York: Prentice-Hall, 1972), pp. 56, 153. Douglas (*Risk and Culture,* p. 131) directly relates the form of organization of the environmental activist groups with their images of disaster.

22. L. Gallonedec, "Man's Dependence on the Earth," *Popular Science Monthly* 53 (May 1898), read in the Congress of Scientific Societies, France.

23. L. Eiseley, *The Star Thrower* (London: Wildwood House, 1978), p. 179.

24. C.G. Jung, *CW* 10, §615.

25. H.B. Hough, in D. Day, *The Doomsday Book of Animals* (London: Ebury Press, 1981), p. 10.

26. J. and P. Phillips, *Victorians at Home and Away* (London: Croom Helm, 1978), p. 18. Nineteenth-century nationalism provided idealized images of unification, coherence and identity, which belied the intense internal fragmentation and conflict. Through imperial imaginings, for example of the Orient, attempts were made to categorize, organize, and control the globe.

27. J. Hillman, "Anima Mundi," *Spring 1982:* 71–93; and R. Sardello, "Taking the Side of Things," *Spring 1984:* 127–35.

28. See R. Nash's comprehensive *Wilderness and the American Mind* (New Haven: Yale University Press, 1973).

29. See A. Portmann's reflections on beauty in nature, in "What Living Form Means to Us," *Spring 1982:* 27–38; also, J. Hillman, "Natural Beauty without Nature," *Spring 1985:* 50–55.

30. See A. Ziegler, "Rousseauian Optimism, Natural Distress, and Dream Research," *Spring 1976:* 54–65; Eiseley, *Star Thrower,* p. 231; Nash, *Wilderness,* p. 165; F. Younghusband, *Wonders of the Himalayas* (1924) (Chandigarh: Abhishek Publications, 1977).

31. S. Larsen, *The Shaman's Doorway* (New York: Harper & Row, 1977), p. 169.

32. The celebration of a generalized, aesthetic and realistic interpretation of natural landscape by Western lovers of wilderness is almost unique. While the Far East also had a long tradition of aesthetic landscape appreciation, it was highly idealized and formalized. Most traditional cultures have revered *specific* geographical *places* (rather than *views*) through the sacred stories that are

associated with them (See M. Eliade, *Sacred and Profane* [New York: Harcourt, Brace and World, 1959]). Traditional cultures generally view sacred sites with a mixture of reverence, awe, caution and fear. Similarly, the unknown land beyond their own territory is usually treated with considerable circumspection. In a previous paper ("Geography of Imagination: Tibet," *Spring 1984:* 195–209), I have discussed the consequences of the Western attempt to remove imaginative paradox from its fantasies of Tibet.

33. See the important essay by L. White Jr., "The Historical Roots of Our Ecological Crisis" in *The Environmental Handbook,* ed. G. de Bell (New York: Ballantine Books, 1970). For a less one-sided analysis that shows a far greater complexity in Western attitudes toward nature, see K. Thomas, *Man and the Natural World: Changing Attitudes in England, 1500–1800* (London: Allen Lane, 1983). See also S. Fox, *John Muir and His Legacy* (Boston: Little Brown & Co., 1981), chap. 11. See also Yi Fu-Tuan's "Discrepancies Between Environmental Attitudes and Behaviour: Examples from Europe and China," for an attempt to go beyond classic texts and correct the naive reification of Eastern ecological practices (P. English and R. Mayfield, *Man, Space and Environment* [New York: Oxford University Press, 1972]).

34. See Nash, *Wilderness,* and also K. Thomas, *Natural World.*

35. E.g., R. Kaplan, "Some Psychological Benefits of an Outdoor Challenge Program," *Environment and Behaviour,* no. 6 (1974); J. Swan, "Sacred Places in Nature," *The Journal of Environmental Education* 15/4 (1983); W. Hammit, "Cognitive Dimensions of Wilderness Solitude," *Environment and Behaviour,* no. 14 (1982).

36. Douglas, *Risk and Culture,* p. 151.

37. Ibid., pp. 131–40 for a discussion of this relation between global fears and global strategies. A thoughtful example of this approach is R. Higgins, *The Seventh Enemy: The Human Factor in the Global Crisis* (London: Hodder and Stoughton, 1978). Kern, for example, relates the emergence of the concept of "open space" with the rise of nineteenth-century imperialism, exploration and the Western domination of the globe (*Culture of Time and Space,* p. 164).

38. Eiseley, *Star Thrower,* p. 262.

39. G. Bachelard, *The Poetics of Reverie* (Boston: Beacon Press, 1971), p. 127.

40. Eiseley, *Star Thrower,* p. 262. These dark murmurs from the environment and nature are also subject to political manipulation, and we have to ask why there are fluctuations in the focus of concern—pollution one year is displaced by energy crisis, over-population by nuclear war and so on.

SKULL'S LANTERN
Psychological Projection and the Magic Lantern

LEE WORTH BAILEY
(Ithaca, New York)

What happens when we turn the theory of projection back upon itself? What unconscious presuppositions are disclosed when we ask "What are we 'projecting' in the theory of projection?" The theory has been subjected to fruitful re-thinking from several points of view.[1] But the root metaphor "projection" itself remains unexplored. Everyday familiarity with cameras and movies has veiled the influence of photographic projection on psychological theory. For the idea of projection in psychology has its imaginative roots in the slide projector and its descendant, the movie projector. The slide projector, known after the Renaissance as the "magic lantern," has decidedly determined Western ways of thinking about the nature of imagination, delusion, and subjectivity. The simple, concrete device of the magic lantern exhibits the philosophical notion that psyche projects purely internal, subjective feelings outward onto a soulless world of objects. It is precisely this problematic subjectivism which calls for a radical revisioning of what we call "projection."

Only rarely has the connection between the magic lantern and the idea of psychic projection been recognized. Jung once remarked, following a dream of a magic lantern hovering above him like a UFO, "I am projected by the magic lantern as C. G. Jung."[2] And Owen Barfield once briefly pointed to the image of the magic lantern hidden behind the Western philosophy of subjectivity. He said that, in the theory of projection, fantasies are seen

---

Lee Worth Bailey, Assistant Professor of Religion at Ithaca College, Ithaca, New York, is the author of "Stereotypes and Captain Hook" in *Rudely Stamp'd: Imaginal Disability and Prejudice* (Washington, D.C.: University Press of America, 1981).

> as sort of an unconscious "projection" of the inner life of feeling and impulse upon an inanimate outer world. Now whenever this excellent word "projection" is spoken or thought, the humble student of meaning . . . makes his bow to an ingenious little machine, [the] "magic lantern." For the curious idea of "projection" which this invention begot in us, is now . . . ubiquitous.[3]

When this inner subject/outer object dichotomy arose into prominence in Western thought, Barfield says, it was imaginally guided by the magic lantern. In the subjectivist theory of projection, "the picture projected —strangely enough—is a picture of a man with a magic lantern."[4] Western subjectivism has been shaped by the root metaphor of the magic lantern, but we are only dimly aware of it.

As a hidden collective image, the magic lantern has misled psychology into imagining that psyche actually does project outward from a contained mind. This image has built into the theory of psychological projection a pervasive distortion: the view that psyche is restricted to the skull. When experienced outside this container, this image teaches, psyche must always be a delusion, an error that must be "withdrawn" back inside. This delimitation of psyche's rightful place, however, distorts the very experience that it is intended to describe. It unnecessarily blocks the disclosure of our participation in Being that the experience now labeled "projection" opens up, as David Holt, Wolfgang Giegerich, and Han Fortmann have pointed out. The root metaphor of the skull as a magic lantern inhibits the task of revisioning soul in the world.

Looking Backwards

Freud's use of projection as a psychoanalytic construct was new, but the idea of projection was a commonplace in late nineteenth-century European psychology. In the same period, the magic lantern was fascinating large crowds with its tricks. Freud first wrote about his theory of projection in 1895, the very same year that the first technically successful movie projector, derived from the magic lantern, was completed by Edison, Lumiere, and others.[5]

Looking backwards from Freud in 1895, we discover that in 1884 Freud's teacher Theodor Meynert used the magic-lantern image in brain anatomy, imagining "projection-fibres" projecting onto the cortex's "tabula rasa."[6] Freud reversed this metaphor, seeing images as if they came from skull's magic lantern, projecting outward as well. In the 1880s

Nietzsche criticized current ideas of psychological projection as being excessively subjectivist.[7] In 1877 Ernest Kapp developed a philosophy claiming that technological devices are unconscious projections of human organs.[8] In 1876 Wilhelm Volkmann took it to be self-evident that perceptions and sensations (*Empfindungen*) are projected into the outer world.[9] In 1870 Hippolyte Taine used the term "projection" in a similar way.[10]

In 1854 Ludwig Feuerbach's scandalous critique of religion was translated into English by George Eliot as *The Essence of Christianity*. Though Feuerbach had not labeled his theory with the word *projicieren*, Eliot introduced the term "projection" into his theory of alienation. Subsequently, her image has wrongly stamped his critique of the Gods as a theory of "projection."[11] And as early as 1851 C. G. Carus used the concept of projection in his psychology of optics.[12]

All this theorizing about "projection" took place in a nineteenth-century context of widespread magic lantern shows traveling across Europe and America. Theaters packed in crowds to view the latest device for projecting mysterious ghosts, angels, or skeletons onto smoke screens, hidden glass panes, or translucent curtains. And of course all this imaginative activity stimulated thinking about projection. Popular magazines revealed these theatrical secrets. But long before Carus and the others began to speak of "projection" in their theories, the image had already become a way of thinking about the very image-making activity of the mind.

The magic lantern developed from Renaissance experiments with a *camera obscura*. Long used for optical investigations, the *camera obscura* was a dark room with a small hole, perhaps containing a lens, that projected the external view onto a white screen. To add a dramatic touch, one Italian experimenter set up images deliberately intended to deceive viewers in the *camera obscura*. He was the playwright Giambattista della Porta, and his 1589 book is the first surviving report of this dramatic effect of a *camera obscura* modified by artificial light and illustration. He described a *camera obscura* whose aperture opened into a second room filled with torchlight. Its illuminated image was mysteriously projected into the adjacent dark room. Della Porta was fascinated with the way it seemed to hang in the middle of the chamber, generating "wonderful mirth . . . fear and terror":

> In a tempestuous night the Image of anything may be represented hanging in the middle of the Chamber, that will terrifie the beholders. Fit the Image

> before the hole, that you desire to make to seem hanging in the Air in another Chamber that is dark; let there be many torches lighted round about. In the middle of the dark Chamber, place a white sheet, or some solid thing, that may receive the Image sent in: for the spectators that see not the sheet, will see the Image hanging in the middle of the Air, very clear, not without fear and terror, especially if the Artificer be ingenious.[13]

From this beginning the story of the magic lantern again and again offered the experience now commonplace in the cinema: spectators in a dark room fascinated by the projection of images, including many technical tricks suggestive of delusions. Cinematic projection suggests that psyche also projects.

A German priest in Rome, Athanasius Kircher (1602–80), also explored the projection of lighted pictures into a *camera obscura*, as is shown in the first edition of his *Ars magna lucis et umbrae* in 1646 (Fig. 1).[14]

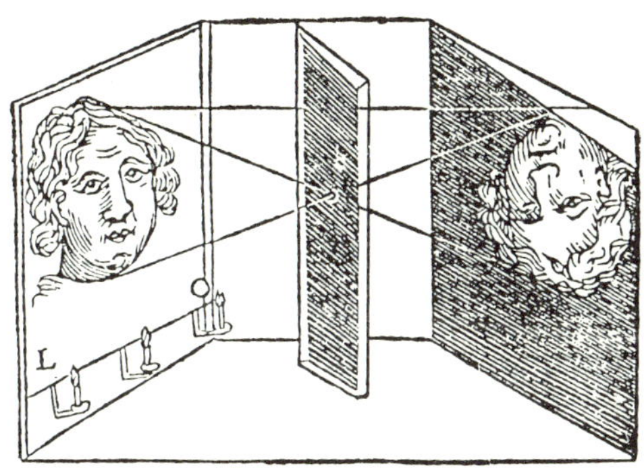

Fig. 1. Projection in a Camera Obscura

In the 1671 second edition of his *Ars magna*, Kircher published the first illustration of the magic lantern proper (Fig. 2).[15] Here Kircher first showed the magic lantern with an artificial light, lens, and painted glass slides. Due to the wide circulation of his book, he was for two centuries wrongly proclaimed the inventor of the magic lantern. Ironically, it seems that Kircher would not have endorsed the magic lantern's subsequent effect on the philosophy of subjectivity. The frontispiece of his *Ars magna* in 1671 was an engraving of a God projecting his rays of light down to earth,

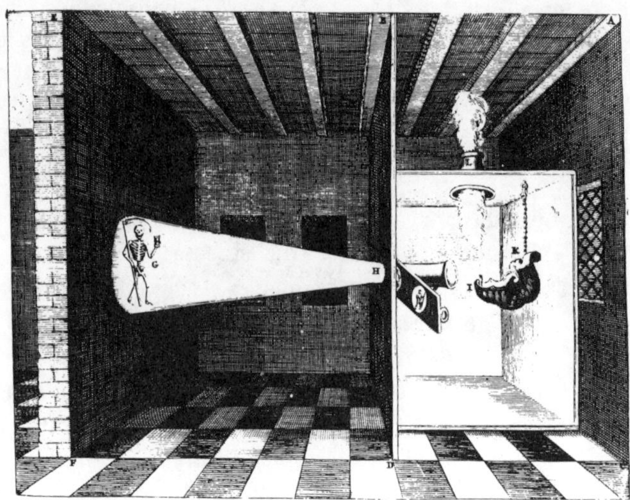

Fig. 2. Kircher's Magic Lantern

which are then directed by human devices (Fig. 3).[16] Any human projection, to him, would originate with God, not in a psychic container.

But before Kircher it was the Dutch physicist Christian Huygens who first recorded the combination of all the elements of the magic lantern around 1659: artificial light, a concave mirror, lenses, and illustrations on a transparent surface (Fig. 4). Nevertheless, Huygens considered the magic lantern a "trifle" and avoided association with this apparatus because it could produce ghostly apparitions.[17]

From its beginnings with Della Porta, Kircher, and Huygens, the magic lantern began its enchantment.[18] And it remains the central image in the theory of projection, where it continues to generate the pervasive root metaphor of psyche, not merely as the passive recipient of visual perceptions as John Locke postulated, but also as the active projector of fantasies. While Locke imagined psyche as a passive *camera obscura*,[19] the magic lantern pictured it as an actively projecting skull.

The first published English record of the magic lantern was that of John Locke's friend William Molyneux. In his *Dioptrica Nova* (1692), he reduced the "Ludicrous or Frightful" images projected from the magic lantern to "no more than . . . representations of outward Objects in a dark Room."[20] The Danish mathematician Thomas Walgensten improved the magic lantern and traveled widely with it, giving demonstrations in

Fig. 3. Kircher's Projecting Divinity

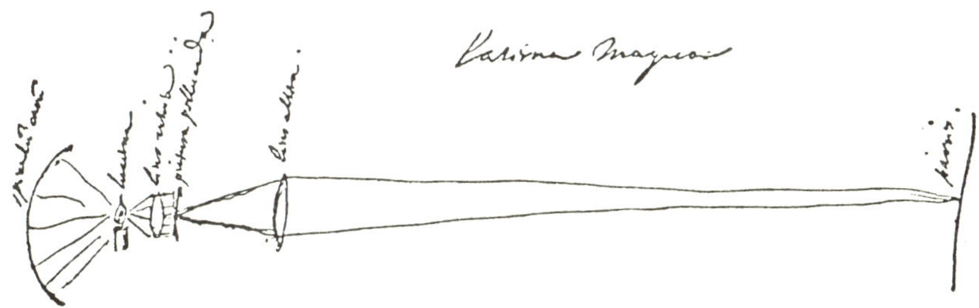

Fig. 4. Early Magic Lantern Technical Diagram

Paris, Rome, and Copenhagen by 1670. Walgensten's apparatus was first called a "scare lantern" (*Schreckenlanterne*) and a "fear lantern" (*Lanterne de Peur*), indicating the fright that it evoked.[21] In France the magic lantern's construction was published in 1643 by the skeptical de Vallement, who lamented, "I know full well that this artifice can be abused at the expense of ignorant and credulous people."[22] These early denigrations of the lantern's imaginative impact lie behind the subsequent practice of envisioning certain psychic phenomena as "nothing but projections."

Today the camera and the projector work as a team, the camera passively recording, the projector, whether slide projector, movie projector, or television tube, actively casting forth the recorded images. The viewer of any of these projections sits in a *camera obscura*, whether a cinema theater or a living room. For the enjoyment or edification of an audience, images from a portable camera's memory are projected into a *camera obscura*. Like Della Porta's astounded spectators, viewers today (though by now accustomed to the magic) participate in the work of the *camera obscura* and the magic lantern as they watch. The literature of the magic lantern reveals the increasing excitement of the latent fantasy that, like the magic lantern, psyche *projects*.

In 1704 John Harris colorfully recorded this excitement in his *Lexicon Technicum*, confirming that the magic lantern was terrifying to many:

> *MAGICK-Lantern*. a little Optick Machine; by the means of which are represented on a Wall, in the Dark, many Phantoms and terrible apparitions, which are taken for the Effect of *Magick* by those that are ignorant of the Secret. This Machine is composed of a Concave Speculum, reflecting the light of a Candle, which passeth through the little Hole of a Tube, at whose End there is fasten'd another Glass: Between these two are successively placed many small Glasses, painted with different Figures, of which the most formidable are always chosen, and such as are most capable of terrifying the Spectators; so that all these Figures may be represented at large on the opposite Wall. . . . They are sold by all Perspective [lens and telescope] makers.[23]

It is because of the magic lantern that "phantoms and terrible apparitions" came to be seen as "projections."

Meanwhile, magicians were using the lanterns for their shows. One clever magician in 1797 shocked Paris at an abandoned Capuchin monastery surrounded by tombs, where spectators entered a dimly lit chamber decorated with skulls, to the sound of mordant music, tolling bells, and thunder. Etienne Robertson then slyly offered to the spectators a show superior to charlatans and their bogus apparitions. He tossed chemicals onto a brazier, producing clouds of smoke. The lights flickered out and, in the darkness, images of skulls, demons, skeletons, and dead heroes appeared, undulating in the smoke-screen. Terrified spectators sank to their knees, drew their swords, or covered their eyes (Fig. 5).[24]

This bone-chilling performance was the public spectacle of Robertson's *Fantasmagorie*, the trickery of the magic lantern. Secretly, the images visible on the smoke screen were projected from behind the audience. Robertson's apparitions created such a sensation that authorities interfered, and he had to be careful in his selection of ghosts. Robertson also developed the means of rear projection onto a translucent sheet which allowed him to move the projector back and dramatically enlarge the image (Fig. 6). Today television tubes operate on the same principle of rear projection.

Fig. 5. Robertson's Fantasmagorie

Fig. 6. Robertson's Rear Projection

Later Robertson took his show to Vienna and St. Petersburg. His *Fantasmagorie* was so successful that the secrets of his magic lanterns were rapidly stolen and similar performances multiplied. In 1801 Paul de Philipstahl took his *Phantasmagoria* to London and then, in 1803, to New York, Cincinnati, and Savannah.[25]

Until 1820 the magic lantern had been fueled by candles and oil. But the invention of the limelight around 1820 introduced a far more powerful illumination which could be directed and focused. In London, John Pepper created dazzling effects using these enhanced projection techniques. During the 1860s he generated quite a stir with "Pepper's Ghost" (Fig. 7). To create this projection, he mounted in front of a stage, at a 45-degree-angle, a huge sheet of glass invisible to the spectators. Then a "ghost" in the pit below, illuminated by a strong lantern, was reflected up and seemed to appear on stage. Variations on this trick spread rapidly to other theaters. The idea that imaginative figures such as ghosts can be demystified and reduced to a "projection" followed these tricks quickly, as popular magazines revealed the stage secrets.

By 1906 the self-proclaimed "Largest Producer in the World" of photographic magic-lantern slides stocked over two million slides for sale.[26] The transformation of the slide projector into the motion picture projector was well underway by the 1890s and, combined with rapid technical developments, soon blossomed into the modern cinema industry. And, of course, all this imaginative activity stimulated more thinking about projection.

Fig. 7. Pepper's Ghost

"Nothing But Projections"

The word "projection" was adopted as an image of mental activity well before the Renaissance. In medieval alchemy it was used as a technical term for the process during which one "metal" showers another with effusions. But the word's connection with the workings of the magic lantern and the lantern's shaping of our ideas of mental activity came later. F. Schumer and J. Zubin found that:

> By the end of the sixteenth century, projection not only denoted the *action* of projection, that is, in throwing or casting forth or forward, but the less tangible mental process, as in the construction of mental projects, scheming and planning.[27]

The magic lantern became a widespread influence behind Western notions of intangible mental processes, such as imagination and thought.

The delusory quality of widespread, mechanically projected images was accepted into eighteenth-century English as an analogy for a deceptive character type. In his picaresque romance *Ferdinand, Count Fathom* (1753), Tobias Smollett, M.D. described "the travelling savoyards who stroll about Europe, amusing ignorant people with the effects of the magic lanthorne."[28] Smollett also frequently used the word "projector" in this novel to describe a posturing fraud engaged in a deceptive scheme of seduction or enrichment.[29] Mechanical or human, a "projector" puts forth a fraudulent image.

By the nineteenth century, visions, demons, and Gods were increasingly being explained away as optical delusions which are, as Freud later put it, "nothing but projections." But well before the scholarly theories of psychology and philosophy, the idea that certain notions are "nothing but projections" appeared in nineteenth-century popular culture, literature, and scientific theory.

By 1808 an English political broadside criticizing civil liberties for Catholics pictured a magic lantern projecting a drawing of the Pope and displayed the lines (Fig. 8):

> What is this spectre of affright
> With which they would delude our sight?
> A shadow thrown upon the wall
> A magic-lanthorn-shew! That's all![30]

This polemical reduction of a religious concern to "nothing but projection'' preceded by over forty years the English translation of Ludwig Feuerbach's epochal book *The Essence of Christianity*, which argued that the Gods are ''nothing but projections.''

Fig. 8. Magic Lantern Political Broadside

In 1826 a French song reveled in the ironies of projected delusions. These verses were printed with an illustration of an itinerant showman offering his traveling magic-lantern show in a *camera obscura* on wheels:

> Whenever deception distorts/Everything that happens here below,/Perhaps the scenes from my *camera obscura* will not offend. Truth in this optical device/To all appears, to all appears./My lantern is truly magic./For a penny you can see it, yes, yes,/For a penny you can see it!/The courtroom where justice/Was always rendered free,/The bedroom of the actress/Where love was always kept away,/And the academic chair/Where slumber never crept. /My lantern is truly magic;/For a penny you can see it, yes, yes,/For a penny you can see it![31]

The magic of the lantern is that it reveals the inner, hidden truth behind delusory, projected images. What a nice idea to borrow!

Charles Dickens included in *The Pickwick Papers* (1837) a chapter in which a church sexton, Gabriel Grub, is terrified in the dark graveyard by what sounds very much like a magic-lantern show. It seems that his friends tricked him with a group of projected ''officious goblins with a perpetual smile upon their faces'' and a display of scenes intended to prick his conscience. One magic lantern specialist comments that "the way these [pictures] occur . . . is reminiscent of a magic lantern twin dissolve unit in operation."[32] Nathaniel Hawthorne also explored the image of psyche projected from a container. In his short story "Fancy's Show Box" (1842), a man imagines his thoughts as if they were projections in a box. "A scheme of guilt," Hawthorne comments, "till it be put in execution, greatly resembles a train of incidents in a projected tale."[33] Here psyche resides only in Fancy's Show Box until acted out, in "projection."

In 1839 Daguerre and others transformed the old *camera obscura* by adding a memory device: light-sensitive plates which could record the inwardly projected image. But long before photography developed, the magic lantern was already suggesting the collective fantasy that, like a darkened box, *a subjective psyche can be imagined as projecting images outward*. Spectators throughout the nineteenth century could go home in awe from a performance by Robertson, Philipstahl, Pepper, and many others, only to read the next day that it was all "nothing but projection." While technical journals such as *The Optical Magic Lantern Journal* explained how magic lanterns worked, an English encyclopedia for young people (1838) highlighted the magic lantern's illusions:

> Here view illusion clad in magic dress,/As here some supernatural agent's aid,/The varied change of aerial forms displayed;/Or some arch demon, lab'ring in the work.[34]

The educational point to young people was: *supernatural forms are nothing but delusions created by the natural magic of projection*.

The Scottish scientist David Brewster (1781–1868) published descriptions of the magic lantern which were soon widely circulated. He criticized what he saw as religious deceptions using such optical projection devices. Brewster's popular *Letters on Natural Magic*, first published in 1832, was immediately translated into German and then reprinted fifteen times in fifty years. Here Brewster revealed the magicians' tricks with the lantern. In the tradition of the natural magic books like Della Porta's, he wanted to reduce all visions and optical illusions to the laws of nature. This reduction was his way of discrediting what he saw as an ancient use of the trickery of the magic lantern for religious deception:

> There is reason to think that [the ancients] employed [lenses and mirrors] to effect the apparition of their gods; and in some of the descriptions of the optical displays which hallowed their ancient temples, we recognize all the transformation of the modern phantasmagoria.[35]

Brewster refers here to his interpretation of accounts from ancient literature which suggest that ancient priests may have used mirrors and other devices to delude their believers with "nothing but projected delusions." For example, he reports that the sixth-century scholar Damascius describes an image in a temple that sounds very similar to Robertson's moving projections:

> There appeared on the wall of the temple a mass of light; it transformed itself, in coming nearer, into a face evidently divine and supernatural, of a severe aspect, but mixed with gentleness, and extremely beautiful. . . . The Alexandrians honored it as Osiris and Adonis.[36]

Brewster similarly interprets an incident in the biography of the sixteenth-century artist Benvenuto Cellini. Witnessing a frightening display of demons appearing one night in the Roman Colosseum, Cellini was terrified but hinted that he was aware of the use of theatrical devices. Going home, his party saw the same demons leaping and skipping before them on the rooftops and on the ground.[37] Brewster concludes that the magic lantern may have been the device behind these demons,[38] but no record confirms the existence of the magic lantern proper until the next century. Cellini's party probably glimpsed shadows of demons painted on an ordinary lantern's glass pane. Such mythic images, Brewster argues, were "a natural system of deception . . . calculated to astonish the beholder, to confound his judgment, to dazzle his senses." Religious images are to him "nothing but projections." Brewster's argument was published widely in England, America, and Germany two decades before the English translation of Feuerbach's book appeared in 1854.

Usually implicit and hidden, the influence of his magic lantern on the notion of the subjective, contained psyche was at times made explicit. In the eighteenth century, a French satire pictured a monkey lecturing to animals, having forgotten to ignite his magic lantern's light.[39] In a similar English satire (1865) of dull, pompous lantern talks, a lecturer also forgot to light the lamp inside his magic lantern. Nevertheless, he proceeded to bore the audience to sleep, blindly giving a slide lecture without the slides:

> None had seen anything at all. . . . He had forgotten at the first/To light the little lamp inside. . . . Staring at nothing, bewildered . . . I think, why, surely *he's* forgot/The light inside *his* lantern-head.[40]

Here appears the explicit comparison of the psyche to a magic lantern: the lantern-head is like a projecting machine. The same fantasy lives on today in a cover illustration for *Psychology Today* magazine, in which psyche is imagined as a projecting machine contained in the skull (Fig. 9).

The magic lantern's influence is not merely interesting; it seriously distorts and reifies the paradoxical relations between the private, hidden, individual psyche and the larger collective world in which it participates, thereby confirming the delusion of a rigid subject/object dichotomy. By picturing psyche as a dark box projecting forth images, the old magic lantern and its theoretical descendants today harden the very divisions between the inner and outer, the soul and the world, and the individual and collective that archetypal psychology has worked so hard to see more paradoxically. The image of "projection" is unnecessary because other terms, such as "attribution," are available. So, useful as the theory of projection has been for disclosing unconscious dynamics, it needs to be thoroughly revisioned, in order to free the phenomenon from the image of projection itself.

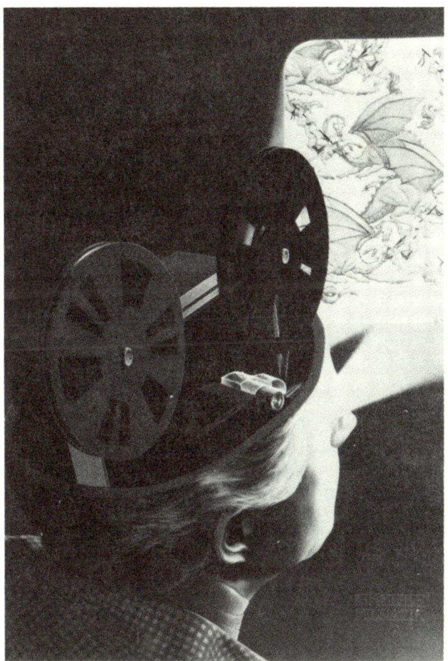

Fig. 9. Psyche imagined as a projecting machine contained in the skull.

1. David Holt, "Projection, Presence, Profession," *Spring 1975*: 130–44. Wolfgang Giegerich, "Der Sprung nach dem Wurf: Uber das Einholen der Projektion und der Ursprung der Psychologie," *Gorgo* 1 (1980): 49–71. Marie-Louise von Franz, *Projection and Recollection in Jungian Psychology*, trans. W. Kennedy (LaSalle, Ill.: Open Court, 1980). Han Fortmann, *Als Ziende de Onzienlijke*, 2 vols. (Hilversum, Netherlands: Gooi en Sticht, 1974).

2. C. G. Jung, *Memories, Dreams, Reflections*, ed. Aniela Jaffé (New York: Random House, 1961), p. 323.

3. Owen Barfield, "Subjective and Objective," in his *Poetic Diction* (Middletown, Conn.: Wesleyan University Press, 1973), p. 203.

4. Ibid., p. 204. See also Barfield's "The Harp and the Camera," in his *The Rediscovery of Meaning* (Middletown, Conn.: Wesleyan University Press, 1977).

5. Sigmund Freud, *The Standard Edition of the Complete Psychological Works of Sigmund Freud*, ed. J. Strachey, 24 vols. (London: Hogarth Press, 1953–74), 1 : 206–12; and Eric Barnouw, *The Magician and the Cinema* (New York: Oxford, 1981), pp. 41–43.

6. Theodor Meynert, *Psychiatry* (New York: Putnam's, 1885), p. 30.

7. Friedrich Nietzsche, *The Will to Power*, ed. W. Kaufmann (New York: Vintage, 1967), pp. 265–73.

8. Ernest Kapp, *Grundlinien einer Philosophie der Technik* (Braunschweig: Westerman, 1877).

9. Wilhelm Volkmann, *Lehrbuch der Psychologie* (1875–76; reprint ed., Cothen: Schultze, 1885), vol. 2, p. 126.

10. Hippolyte Taine, *On Intelligence* (New York: Holt, 1875), vol. 2, p. 90.

11. Ludwig Feuerbach, *Gesammelte Werke*, ed. W. Schuffenhauer, vol. 5, *Das Wesen des Christentums* (Berlin: Akadamie-Verlag, 1973), pp. 44, 71, 360, 377. In English: *The Essence of Christianity* (New York: Harper & Row, 1957), pp. 11, 29–30, 313, 226.

12. C. G. Carus, *Physis* (Stuttgart, 1851), p. 415.

13. Giambattista Della Porta, *Natural Magic* (1589; English translation: London, 1658), bk. 17, ch. 7.

14. Athanasius Kircher, *Ars magna lucis et umbrae* (Rome, 1646), pp. 121–22 and 907–16.

15. Kircher, *Ars magna lucis et umbrae*, 2d ed. (1671), bk. 10 (*Magia*), pt. 3, pp. 768–69.

16. Joscelyn Godwin, *Athanasius Kircher* (New York: Thames and Hudson, 1979), p. 78.

17. Christian Huygens, *Oeuvres Completes* (Le Havre: Nijhoff, 1888–1967), vol. 4, pp. 102–11, 125, 197, and 269. F.P. Liesegang, "Christian Huygens und die Zauberlanterne," *Central-Zeitung für Optik und Meckanik* 50/9 (May 5, 1929): 167-69. Also W. A. Wagenaar, "The True Inventor of the Magic Lantern," *Janus* 66 (1979): 193–207.

18. Barnouw, *Magician and Cinema*.

19. John Locke, *Essay on Human Understanding*, ed. John Yolton (New York: Dutton, 1976), bk. 2, ch. 11, §17.

20. William Molyneux, *Dioptrica* (London, 1692), p. 183, prop. 59.

21. J.M. Eder, "The Invention of Projection Apparatus," in his *History of Photography* (New York: Columbia, 1905), p. 52.

22. R.G. Morris, "The Magic Lantern in 1693," *New Magic Lantern Journal* 1/1 (April 1978): 3. M. de Vallement, *La Physique Occult* (Amsterdam, 1693), pp. 420–21.

23. John Harris, *Lexicon Technicum* (London, 1704), Magick-Lantern.

24. Barnouw, *Magician and Cinema*, p. 19.

25. Ibid., pp. 22–23.

26. Robert McMillan, "James Bamforth," *New Magic Lantern Journal* 1/2 (February 1979): 14–16.

27. F. Schumer and J. Zubin, "Projective Techniques," *Encyclopedia of Psychology* (New York: Seabury, 1979), p. 833.

28. Tobias Smollett, *Ferdinand, Count Fathom* (1753; reprint ed., New York: Greenberg, 1926), ch. 58, p. 384.

29. Ibid., pp. 174, 196, 212, 214, 272, and 398.

30. John Barnes, *Optical Projection* (St. Ives, Cornwall, England: Barnes Museum of Cinematography, 1970), p. 40.

31. "La Lanterne Magique," lyrics by A. Naudét (Paris, 1826); in *New Magic Lantern Journal* 2/2 (January 1982): 8–9.

32. William Main, "Did Charles Dickens Write Tales Specifically for the Magic Lantern?" *The Magic Lantern Society of Great Britain Newsletter* 6 (May 1982): 2. Charles Dickens, *The Works of Charles Dickens*, vol. 6, *The Pickwick Papers* (New York: Appleton, 1871), ch. 29, pp. 164–68.

33. Nathaniel Hawthorne, "Fancy's Show Box," in *Twice-Told Tales* (1842; reprint ed., New York: Lupton, 1943), pp. 265–73.

34. R. G. Morris, "Victorian Youth and the Magic Lantern," *New Magic Lantern Journal* 2/1 (January 1981): 2–3. W. Montague, *The Youth's Encyclopedia* (Emans, 1838).

35. David Brewster, *Letters on Natural Magic* (New York: Harper, 1832), p. 17.

36. Ibid., p. 69. Damascius, "Vie d'Isidore," in Proclus, *Commentaire sur la Parmenides* (1903; reprint ed., Frankfurt: Minerva, 1962), vol. 3, p. 301.

37. Benvenuto Cellini, *Autobiography*, trans. J. Symonds, Harvard Classics 31 (New York: Collier, 1960), pp. 127–30.

38. Brewster, *Letters*, pp. 70–77.

39. Jean-Pierre Florian, "Le Singe qui Montre la Lanterne magique," in his *Fables* (Paris: Houdaille, 1843?), vol. 2, p. 7.

40. Darcy Thompson, *Fun and Earnest* (London: Griffin, 1865); in *New Magic Lantern Journal* 2/1 (January 1981): 11.

List of Illustrations

Fig. 1. *Projection in a Camera Obscura*. 1646. Athanasius Kircher. *Ars magna lucis et umbrae*. Rome, 1646, p. 121.

Fig. 2. *Kircher's Magic Lantern*. 1671. Athanasius Kircher. *Ars magna lucis et umbrae*, p. 769.

Fig. 3. *Kircher's Projecting Divinity*. 1671. Athanasius Kircher. *Ars magna lucis et umbrae*. Frontispiece.

Fig. 4. *Early Magic Lantern Technical Diagram*. 1694. Christian Huygens. *Dioptrique*, pp. 2, 786. Volume 13 of his *Oeuvres Completes*. Le Havre: Nijhoff, 1888–1967.

Fig. 5. *Robertson's Fantasmagorie*. c. 1797. Etienne Robertson. *Memoires recreatifs scientifiques et anecdotiques*. Paris, 1831. Frontispiece.

Fig. 6. *Robertson's Rear Projection*. 1849. *Le Magasin Pittoresque*. Paris, 1849.

Fig. 7. *Pepper's Ghost*. 1860s. At London's Royal Polytechnic. In *The Magic Lantern: How to Buy and How to use it and How to Raise a Ghost*, by "A Mere Phantom." 1880.

Fig. 8. *Magic Lantern Political Broadside*. 1808. Barnes Museum of Cinematography, St. Ives, Cornwall, England.

Fig. 9. *Psyche imagined as a projecting machine contained in the skull*. Kimmerle Milnazik. *Psychology Today* magazine cover, September 1980. Used by permission of the artist.

JUNG, HEGEL, AND THE SUBJECTIVE UNIVERSE

BARBARA ECKMAN
(Philadelphia)

Since this is to be an essay on subjectivity which is external to human persons, it may be well at the outset to make clear what exactly I mean by "subjectivity." I do not intend a (merely) private, personal, or idiosyncratic viewpoint: "my innermost subjectivity." Nor do I refer to the quality inhering in such viewpoints as when a husband tells his wife, "You're too subjective—why can't you be objective for a change, like me?" Subjectivity as I am using the term, and as Hegel uses it, entails the quality of essential, living integrity. A subject possesses autonomy, takes initiative, and is both self-determined and self-determining.[1] As Hegel puts it, a subject is that which is "actual only insofar as it is the movement of positing itself, or the mediation between a self and its development into something different."[2] In Walter Kaufmann's words, "what Hegel means by a subject is that which makes itself what it becomes."[3]

The sense of isolation and emptiness characteristic of our civilization springs from the Enlightenment view that the correlation between events and meaning resides only in the human mind and not in the extra-human world (cf. *CW* 8, §928). If meaning and purposefulness seem to reside in extra-human events and extra-conscious personages, it is on the Enlightenment view simply because we have placed it there, by projection—have merely lent the chaotic, meaningless external world a reasonable, orderly appearance by superimposing human reason upon it. From this perspective, nothing extra-human actually addresses us;

Barbara Eckman does psychotherapy with emotionally disturbed children in suburban Philadelphia and occasionally teaches Religion and Psychology at the University of Pennsylvania. Other writings include psychological interpretations of computers and moon colonization fantasies.

nothing outside us ever speaks first; all is chaos, meaningless, until we utter the Jehovan Word, "Let there be."

Jungian, like Hegelian, thought urges us beyond this "Enlightened" concept of projection into an authentic relation with the extra-human world, a world of autonomous life which arouses and demands a loving response. But Jung and Jungians after him, for the most part, base their psychologizing on a Kantian epistemology, which is in the Enlightenment tradition.[4] This produces a fundamental problem in Jungian thought. By turning from Kant to Hegel, we may discover a new philosophical context for the Jungian conception of a subjective universe.

I. Kant and the Subject-Object Dichotomy

One of Kant's great achievements was to differentiate the subject from the object of knowledge, the operation that Jung calls the dissolution of the *participation mystique*—the primitive, unconscious state of undifferentiated unity between psyche and world, the boundaries between mind and matter not yet clearly drawn, and mind assumed to exercise magical power over matter. According to Kant, all knowledge of the material world contains a "subjective" component provided by the nature of the mind, as well as an "objective" component provided by the objects of knowledge (what Kant called the things in themselves, *die Dinge an sich*). Because of this subjective component, it is impossible to apprehend things as they are in themselves; they may be known only as *phenomena*—as they appear to us.

Kant's bifurcating the subject and object of knowledge, while necessary for the development of consciousness, nevertheless serves to exacerbate the problem we wish to solve. It opens an impassable chasm between the living, active, knowing human subject and the passive, inanimate object of knowledge. Even God in Kant's scheme is not capable of bursting through the boundaries and limitations of human thought the way he can in, say, Augustine's, but must stand still, submit to analysis by critical reason, and wait to be posited by faith as the ground of our concept of human moral freedom. God in Kant's thought functions not as subject but as object—and to this extent at least is not alive but dead! While this portrayal of God is not in the least absurd in the context of Enlightenment philosophy, from the Jungian perspective of the God-image as an autonomous unconscious personality, it has an inauthentic if not neurotic ring.[5]

Kant's thought cannot provide a philosophical context in which to overcome the modern alienation between humanity and world, for it was Kant's philosophy that actually secured the foundation for the problem in the first place. Hegel, though influenced by Kant, presents a more unitary, teleological, and subjective vision of the cosmos. Despite Jung's personal antipathy to Hegel's writing,[6] Jungian thought's ultimate aims and vision of the universe are actually much closer to Hegel than to Kant, and find in Hegel's thought a more fitting philosophical ally.

II. Extra-Human Subjectivity in Hegel

Kant adopts the Cartesian separation of subject and object of knowledge. Hegel has no intention of denying or ignoring this distinction, which would return him to a naive state of *participation mystique* between knower and known. As Hans-Georg Gadamer puts it,[7] he seeks to overcome the apparent contradiction between the philosophies of Descartes and the Greeks (chiefly Plato and Aristotle) by finding a third position, a "higher unity," which embraces and thereby reconciles the two opposing viewpoints.[8] Hegel would retain, among other things, the "Enlightened" reflection which Kant prizes so highly. But according to Hegel, analysis is always followed by synthesis;[9] tearing apart leads to reuniting. The Kantian subject–object split must eventually be seen through and overcome, so that the difference between subject and object is recognized as being in actuality "no difference."[10] Consequently, one of the elements of Greek philosophy which Hegel values and seeks to retain in his philosophy is the doctrine of *nous.*

Generally speaking, the Platonic and Aristotelian doctrine of *nous,* the mind or reason, later developed as *logos* in Stoicism and Neoplatonism, posited reason as universal. *Nous* gave order and rationality to the universe and made it a *kosmos.* All that was knowable by means of human reason, the *nous* in us, was itself *nous;* "only like could know like." Thus any act of "divining" rationality or order in the sensible world in the fields of geometry or philosophy consisted in *nous* knowing itself. *Nous* in the philosopher, for instance, recognized *nous* in nature as the laws of geometry or the form of the beautiful.

Hegel explicitly affirms the doctrines of *nous* and *logos* in his discussion of the teleological argument for God's existence (purposefulness in nature bespeaks a purposeful creator), in which he attempts to rehabilitate this traditional proof in the wake of the Kantian critique:

> The world is a harmonious whole, an organic life which is determined in accordance with ends. It was this which the ancients held to be *nous,* and, taken in a more extended signification, this life was also called the world-soul, the *logos*. . . . When we grasp the fact of life in its true nature, it is seen to be one principle, one organic life of the universe, one living system. All that is, simply constitutes the organs of the one subject. The planets which revolve round the sun are simply the giant members of this one system. Regarded in this fashion, the universe is not an aggregate of many accidents existing in a relation of indifference, but is a system endowed with life.[11]

This Hegelian conception, following Plato and Neoplatonic thought, of the world as "one living system" or "one subject" leads to a view of the world and humanity's place in it which is so drastically different from the Kantian–Enlightenment view that it is difficult to describe vividly enough. From the "Enlightened" perspective, human knowing beings are fundamentally alien to the world they inhabit. The night in its blackness presses upon them and threatens to overwhelm them with its vacuity, devoid of the consciousness and subjectivity which set them apart and which they must use to understand, control, and order the alien world around them. For Hegel humanity is not fundamentally alien to the world; the night is peopled with beings which cluster around our dwellings, pressing their faces to the windows, regarding us with strange, nonhuman curiosity in their bright yellow eyes.[12]

III. Extra-Human Subjectivity in Jung: The Psychoid Archetype

On the basis of his clinical experience, Jung was convinced that the only remedy for the modern spiritual problem was to reestablish contact and even communication with the deep sources of psychic life: the archetypes of the collective unconscious. Contact with these is more than "getting to know oneself," or "getting in touch with one's feelings," or even reconstructing or reliving primal childhood experiences constitutive of one's abiding psychological conformation. In contacting the archetypal dimension of the psyche, one meets trans-personal, primitive, peculiarly and universally human modes of "apprehension"—of experiencing life.[13] Further, archetypes typically are experienced as autonomous personalities;[14] Jung even recommends carrying on conversations with them in active imagination as a technique for tapping their wisdom. Such descriptions and techniques illustrate how much Jung thought of the archetypes as living personalities—purposive, active, and intelligent, if sometimes also rather primitive and archaic (that is, underdeveloped as compared with conscious ego functioning).[15]

Such is Jung's theory of archetypes in his middle years, the years that produced the "mainstream" Jungian archetypal theory, that most widely identified as "Jungian" by the larger intellectual community. The principal data used by Jung to generate this theory were the following: schizophrenic delusions; dreams, fantasies, and artistic productions of "normal" patients; and "collective dreams" such as mythic stories and fairy tales. But during the later part of Jung's life, from about 1946[16] until his death in 1961, Jung focused more intently than ever before on the study of parapsychological phenomena.[17] This late period in Jung's thought produced his theory of synchronicity and the related redefinition of the archetype as "psychoid," not psychic, reality.

The archetype as psych*ic* reality only was not able to explain apparently "meaningful coincidences" between psychic states and external events—events contravening statistical probability.[18] Jung found unconvincing attempts to provide causal explanations of such phenomena, for example as evidence of extrasensory perception (ESP) or psychokinetic (PK) abilities, in which either the mind caused events to happen or events produced mental images at a distance. To explain coincidence between events inside and outside psychic reality for which a causal explanation was lacking, Jung appealed to a reality underlying both material and psychic reality: the "psychoid" archetype or archetype per se, as distinguished from the psychic archetypal image or symbol.

The relation of the psychoid to material and psychic reality is given in Jung's two major essays developing the psychoid theory, "On the Nature of the Psyche" (1946; *CW* 8, §§343–442) and "Synchronicity: An Acausal Connecting Principle" (1952; *CW* 8, §§816–968). Jung postulated the psychoid archetypes (*CW* 8, §§417–18) as "transcendent" or "transcendental" factors,[19] "irrepresentable" in themselves, represented only by their images in psychic or material reality. These psychoid factors inhere in and structure both instinct and (psychic) archetypal symbols. They are, Jung insists, a *description* of the fact (and neither the *cause* nor the *explanation* of the fact) that similar structures seem isolable in both the material and the psychic realms. In the case of so-called synchronistic events or "meaningful coincidences," seemingly inanimate physical reality exhibits characteristics similar to psychic states. By calling these coincidences "meaningful," Jung means not a projection of psychic intention onto dead matter, but an "equivalence or conformity" (*CW* 8, §942, n. 71) between two parts of the same whole which is structured everywhere and throughout by one inherent order:[20]

> Since psyche and matter are contained in one and the same world, and moreover are in continuous contact with one another and ultimately rest on irrepresentable, transcendental factors, it is not only possible but fairly probable, even, that psyche and matter are two different aspects of one and the same thing. The synchronicity phenomena point, it seems to me, in this direction, for they show that the nonpsychic can behave like the psychic, and vice versa, without there being any causal connection between them. (*CW* 8, §418)

But what are the characteristics of this "order" which underlies the entire cosmos, whether material or psychic? Jung compares this inherent order with the Eastern concept of Tao, translated by Richard Wilhelm as "meaning" (*Sinn; CW* 8, §917). According to Wilhelm and Jung, reality is "conceptually knowable" in Taoism because "there is in all things a latent 'rationality,' " which underlies meaningful coincidence (*CW* 8, §922). "The same *living reality* [my emphasis] is expressing itself in the psychic state as in the physical" (*CW* 8, §865). Jung also mentions the following Western, premodern parallels: medieval "correspondence" theory (*CW* 8, §924); the Classical idea of the "sympathy of all things" (*CW* 8, §927); Pico's conception of the cosmos as a living body (*CW* 8, §927); Neoplatonic *archai,* or first principles upholding cosmic order;[21] and the concept of the World Soul that penetrates all things, the foundation of Kepler's observation that the natural or living power in bodies has a "divine similitude" (*CW* 8, §936).

Jung also points out that "a suggestion of a self-subsistent meaning" in the material world is found in the dreams of contemporary subjects (*CW* 8, §945) involving, *inter alia,* representations of human heads which have been carved into the living rock without the help of human artifice. According to Jung, such dreams describe a "meaningful coincidence of an absolutely natural product with a human idea apparently independent of it" (*CW* 8, §946). Jung could have gone even further and interpreted such dreams as testimonies to the "psyche-like" (psychoid) life or subjectivity inherent even in inorganic rock, in the manner in which Hegel saw the physical universe as a living self. But Jung never went this far—probably due to his lingering Kantian assumptions.

Yet the psychoid archetypes, especially when seen as developing out of the earlier (and never abandoned) conception of archetypes as unconscious personalities, appear to possess for Jung the characteristics of subjectivity and life (but *not* of consciousness). The Kantian bifurcation of spirit and matter, subject and object, is of little help in providing a philosophical context for the psychoid archetypes. But Hegel's vision of the subjectivity of being itself, which underlies and enlivens the material universe as well as human subjects, seems perfectly suited to this task.

IV. The Subjective Universe and the Conscious-Unconscious Dichotomy

If Jung's conception of psychoid archetypes overcomes the Kantian bifurcation of subject and object, it fails to address another dichotomy remaining in Jungian thought, the split between the conscious and the unconscious psyche. Might this remaining vestige of "Enlightened" dualism be overcome?

The understanding of the conscious–unconscious division which originated in the pioneering speculations of Freud, Jung, and their colleagues early in this century is well known. Consciousness is commonly understood as the mental activity and realm of which we either are already aware or could easily become aware simply by refocusing our attention.[22] "The unconscious" commonly refers to the (relatively) unknown realm from which dreams and other fantasy material are supposed to come, but of which we have no immediate awareness.[23]

Jung's later thought offers an alternative view which partially obscures this neat distinction between conscious and unconscious psyche, the alchemical concept of the *scintillae* of the *lumen naturae*, or multiple centers of consciousness *within* the unconscious (*CW* 8, §§388–89). This *conscious* quality of "unconscious" centers enables dream and fantasy figures to respond intelligently to changing environmental conditions, within the human psyche or in the external world, and to carry on conversations with human consciousness in active imagination. But precisely because the unconscious is unknown, we cannot validly assume that these "unconscious" centers of consciousness are located inside the human head or the individual psyche at all. From a notion of conscious and unconscious psyche located in individual human brains, we can move to a conception of multiple consciousnesses, some human, others inhuman; some located within human psyches, others located outside, in nature and even elsewhere; some humanly personal and immediately known, others inhuman, impersonal, and only indirectly known by human consciousness. This imagining of multiple consciousnesses in the world, released from their Enlightened imprisonment within human brains, brings threat as well as promise. It may fragment the world and our experience far more radically than the Kantian or intrapsychic subject–object split.

Such a radical subject–subject alienation is inevitable only if these multiple (human and other) consciousnesses are viewed as essentially separate and unrelated to each other, monadic egos chanting "cogito, ergo sum." Instead, we may extend Hegel's conception of the World Soul and imagine all the centers of consciousness, human and inhuman, "conscious" and "unconscious," personal and impersonal, collective and individual, as "organs" of one living system, Universal Psyche (in Hegelian terms, "World Spirit"). As parts of the same organism, the multiple centers of consciousness are no longer alienated and irreconcilably separate from one another, but rather "belong to" one another and to Psyche. As Hegel puts it, their difference is "no difference" because merely an "inner difference" within a single, self-same organism.[24]

V. The Heart of the Matter: Subjectivity and Love

Enlightenment knowledge is knowledge by a rational, self-determining subject of an object, a dead entity without autonomy or integrity. One feels no "brotherhood of being" with an object, and so feels justified in treating it in any way one pleases.[25] For Hegel, Plato, and Jung, knowledge is knowledge of one subject by another—a calling of deep unto deep, an immediate recognition of subjectivity in the other and a reflective recognition and affirmation of the kinship between oneself and the other. As Hegel puts it, "life senses life."[26]

Hegel provides a further hint about the real nature of knowledge among subjects in the early fragment "On Love":

> True union, or love proper, exists only between living beings who are alike in power and thus in one another's eyes living beings from every point of view; in no respect is either dead for the other.[27]

Knowledge of one subject by another is love.[28] "Love is a sensing of something living."[29] The love aroused and demanded by the encounter with another subject has at least two senses: that of acknowledging and valuing, as much as oneself, the other's autonomy, initiative, right to self-determination, and integrity; and also the erotic sense—the magnetic, uniting force which forms a bond between "likes." In a Platonic universe, "like knows like" means "like loves like."

Ultimately, the knowledge that one subject has of another is love. This obtains not only between persons, but also between human and extra-human subjects—be they one's inner voices (the "little people" of the psyche) or the everyday "objects" of the material world, enlivened by the *anima mundi.*

In developing his conception of the psychoid basis of all reality, whether material or psychic, Jung effectively solved the modern spiritual problem of alienation. But the Cartesian–Kantian worldview which gave birth to the problem still remains, in Western culture in general and in Jungian thought in particular. Our isolation and loneliness in the midst of an alien world will not be overcome until the "Enlightened" underpinnings of Jung's thought are seen through, and Jung's Hegelianism *malgre lui* affirmed.

1. Cf. Plato, *Phaedrus* 245e, trans. R. Hackforth, in *Plato's Phaedrus* (Cambridge and New York, 1952), reprinted in Edith Hamilton and Huntington Cairns, eds., *The Collected Dialogues of Plato Including the Letters,* Bollingen Series 71 (Princeton: Princeton University Press, 1961), p. 493:

> And now that we have seen that that which is moved by itself is immortal, we shall feel no scruple in affirming that precisely that is the essence and definition of soul, to wit, self-motion. Any body that has an external source of motion is soulless, but a body deriving its motion from a source within itself is animate or *besouled,* which implies that the nature of soul is what has been said.

If each individual object in the universe is not "besouled," the universe itself, taken as one enormous "body," is.

2. G. W. F. Hegel, *The Phenomenology of Spirit,* §18, trans. Walter Kaufmann, in *Hegel: Texts and Commentary* (Notre Dame: University of Notre Dame Press, 1965), p. 28. Unless otherwise stated, all references to the *Phenomenology* are to the A. V. Miller translation (Oxford: Oxford University Press, 1977). References to Kaufmann's translation will give Miller's paragraph numbers as well as Kaufmann's page numbers.

3. Kaufmann, *Hegel: Texts and Commentary,* p. 31.

4. Jung claims that his archetypes correspond formally to Kant's a priori forms and categories, in that they are present in the human mind prior to the individual's experience, and make apprehension of experience possible. Whether Jung used Kant in a way Kant himself would approve is beyond the scope of this discussion. I suspect that Jung understood neither the motivation and purpose of Kant's system nor the precise meaning and function of the Kantian terms *noumenon* and *phenomenon* in their relation to knowledge and faith. There have been several studies of the similarities and parallels between the systems of Jung and Kant; for example, Eugen Bär, "Archetypes and Ideas: Jung and Kant," *Philosophy Today* 20 (1976): 114–23; Charles E. Scott, "Archetypes and Consciousness," *Idealistic Studies* 7 (1977): 28–49; Marian L. Pauson, "C. G. Jung and the *A Priori,*" *Tulane Studies in Philosophy* 18 (1969): 93–103; Anneliese A. Pontius, "Philosophischer Überblick über die Entwicklung von Jungs Begriff des objektiv Psychischen,"

Zeitschrift für Analytische Psychologie 2 (1971): 69–78. The question whether Jung is true to Kant is rarely addressed. A notable and lively exception (which confirms my suspicions regarding Jung's understanding of Kant) is Stephanie de Voogd's "C. G. Jung: Psychologist of the Future, 'Philosopher' of the Past," *Spring 1977:* 175–82. In this vivid and persuasive paper, de Voogd argues that, in calling archetypal psychic structures *noumena* and archetypal psychic images *phenomena,* Jung misuses Kant's terms, with the result that he actually begins to "dissolve" Kant's world view. On Jung's dissolution of Kant's phenomenon–noumenon distinction, see also Roberts Avens, *Imagination Is Reality* (Dallas: Spring Publications, 1980), p. 93.

5. Cf. *CW* 13, §51, regarding "the reason why our time has become so utterly godless and profane":

> Our true religion is a monotheism of consciousness, a possession by it, coupled with a fanatical denial of the existence of fragmentary autonomous systems. . . . Our time has committed a fatal error; we believe we can criticize the facts of religion intellectually. Like Laplace [and Kant?], we think God is a hypothesis that can be subjected to intellectual treatment, to be affirmed or denied. We completely forget that the reason mankind believes in the "daemon" has nothing whatever to do with external factors, but is simply due to a naive awareness of the tremendous inner effect of autonomous fragmentary systems.

Cf. also the lecture to the student *Zofingiaverein,* called "Thoughts on the Interpretation of Christianity, with Reference to the Theory of Albrecht Ritschl," *CW Suppl.* A, §§248–49.

6. Jung's feelings always run high when he speaks of Hegel. The following are typical of Jung's remarks on Hegel:

> The victory of Hegel over Kant dealt the gravest blow to reason and to the further development of the German and, ultimately, of the European mind, all the more dangerous as Hegel was a psychologist in disguise who projected great truths out of the subjective sphere into a cosmos he himself had created. (*CW* 8, §358)

> A philosophy like Hegel's is a self-revelation of the psychic background and, philosophically, a presumption. Psychologically, it amounts to an invasion by the unconscious. The peculiar high-flown language Hegel uses bears out this view; it is reminiscent of the megalomaniac language of schizophrenics, who use terrific spellbinding words to reduce the transcendent to subjective form, to give banalities the charm of novelty, or pass off commonplaces as searching wisdom. So bombastic a terminology is a symptom of weakness, ineptitude, and lack of substance. But that does not prevent the latest German philosophy [probably Heidegger—see *C. G. Jung: Letters,* Bollingen Series 95 (Princeton: Princeton University Press, 1973, 1976), vol. 1, p. 331 and vol. 2, p. 501] from using the same crackpot power-words and pretending that it is not unintentional psychology. (*CW* 8, §360)

> Of the nineteenth-century philosophers, Hegel put me off by his language, as arrogant as it was laborious; I regarded him with downright mistrust. He seemed to me like a man who was caged in the edifice of his own words and was pompously gesticulating in his prison. (*Memories, Dreams, Reflections* [New York: Random House, 1961, 1962, 1963], p. 69)

Hegel's language in the *Phenomenology* seems to me less "bombastic" than simply unclear, as if he lacked the razor-sharp analytical skills necessary to craft his profound insights into straightforward declarative sentences, and instead was blessed with tremendous intuitive capabilities. Its difficulty does not seem intended to intimidate or bewitch, and it is rarely arrogant—though perhaps the all-encompassing character of his system and his quest to overthrow the traditional logic of noncontradiction do reflect something of *puer* as well as *senex*. Hegel's language remains true to the unconscious origins of his intuitive insight; he does not seek to offend or master his muse by means of penetratingly precise analytical language. This is not the language of the psychotic, but simply that of intuitive sensitivity.

7. Hans-Georg Gadamer, "Hegel and the Dialectic of the Ancient Philosophers," in *Hegel's Dialectic: Five Hermeneutical Studies,* trans. P. Christopher Smith (New Haven: Yale University Press, 1976), pp. 5–34.

8. This process of simultaneous preservation and destruction of opposition by finding a "higher unity" is the famous Hegelian *Aufhebung,* often translated "sublation" or "supersession," the movement which propels his dialectic. For the Jungian equivalent of the Hegelian *Aufhebung,* see the definition of "symbol" in *Psychological Types* (*CW* 6, §§814–29, esp. 824ff.), and the discussion in the "Commentary on *The Secret of the Golden Flower*" on "outgrowing" a conflict by reaching a "higher level" of consciousness (*CW* 13, §17f.). The similarity of the "Golden Flower" text to Hegel's (and Nietzsche's) dialectic has been noted by Walter Kaufmann in his *Nietzsche* (Princeton: Princeton University Press, 1974), p. 82 n.10.

9. See Mark C. Taylor, "*Itinerarium Mentis in Deum:* Hegel's Proofs of God's Existence," in *Deconstructing Theology,* American Academy of Religion Studies in Religion 28 (New York: Crossroad and Chico, California: Scholars Press, 1982), pp. 25–27.

10. Hegel, *Phenomenology of Spirit,* §161. Such a difference is "no difference" because it is a "difference of what is self-same," that is, an "inner difference" within a larger unity (§§160–61).

11. G. W. F. Hegel, "Amplification of the Teleological Proof in the Lectures on the Philosophy of Religion Delivered in the Summer of 1831," in *Lectures on the Philosophy of Religion Together with a Work on the Proofs of the Existence of God,* vol. 3, trans. E. B. Speirs and J. Burdon Sanderson (London: Kegan Paul, 1895), pp. 342–43. Life is also a central concern to Hegel in the *Phenomenology of Spirit,* §§162, 169–76.

On the world as a living system, cf. Jung's paraphrase of Pico's position (a forerunner of his synchronicity concept) in *CW* 8, §927:

> For him the world is *one* being, a visible God, in which everything is naturally arranged from the very beginning like the parts of a living organism. . . . Just as in a living body the different parts work in harmony and are meaningfully adjusted to one another, so events in the world stand in a meaningful relationship which cannot be derived from any immanent causality. The reason for this is that in either case the behaviour of the parts depends on a central control which is supraordinate to them.

12. Cf. Maurice Sendak, *Where the Wild Things Are* (New York: Harper & Row, 1963).

13. C. G. Jung, "Instinct and the Unconscious," *CW* 8, §280.

14. See, for example, *Memories,* where Jung describes his encounter with the unconscious in the form of his "No. 2" personality (pp. 89ff.) and of Elijah, Salome, and Philemon (pp. 181–85). Also "Golden Flower," *CW* 13, §§49–51, 61–62.

15. See, for example, *CW* 7, §§219, 235, 240; *CW* 12, §12.

16. C. G. Jung, "On the Nature of the Psyche," *CW* 8, §§343–442.

17. Although he did not write extensively on the subject in the intervening years, parapsychology had fascinated Jung since his M.D. dissertation in 1902, "On the Psychology and Pathology of So-Called Occult Phenomena" (*CW* 1), and was an important issue, though one often overlooked by commentators, in his break with Freud.

18. For examples of synchronistic phenomena, see Jung's encounters with the fish symbol (*CW* 8, §970) and the scarab symbol (*CW* 8, §982).

19. As can be seen from the context, Jung means a reality "transcendent" to both psychic *and*

physical reality. Hence more is involved than Kantian "transcendental" categories of thought which precede and make possible each and every human psyche's *experience*. Like the archetypes of Jung's middle period, conceived as inherited psychic structures or predispositions, Kantian categories in Jung's terms would be simply a special type of psychic reality, and not "transcendent realities." Unlike Kant, Jung uses "transcendent" and "transcendental" interchangeably.

20. Jung sees the development of the concept of synchronicity as parallel to the attempts of modern physics to attain a unified field theory, or "unitary idea of being" reminiscent of the unitary ontologies of Plato and Hegel, and radically dissimilar to Kant's dualist ontology. In his movement to positing a unitary, psychoid structure underlying the material and psychic worlds, Jung has shifted, without realizing it, from the Kantian position to a more Platonic or Hegelian position:

> If the latest conclusions of science are coming nearer and nearer to a unitary idea of being . . . it seems to show that there is some possibility of getting rid of the incommensurability between the observed and the observer. The result, in that case, would be a unity of being which would have to be expressed in terms of a new conceptual language—a "neutral language," as W. Pauli once called it. (*CW* 8, §960)

21. That NeoPlatonism underlies Jung's view of soul has been highlighted by James Hillman and further advanced in the literature of archetypal psychology. See, for example, James Hillman, "Plotino, Ficino, and Vico as Precursors of Archetypal Psychology," in *Loose Ends: Primary Papers in Archetypal Psychology* (Spring Publications, 1975), pp. 146–69; and Roberts Avens, *Imaginal Body: Para-Jungian Reflections on Soul, Imagination and Death* (Washington, D.C.: University Press of America, 1982); compare also James Hillman, "*Anima Mundi:* The Return of the Soul to the World," *Spring 1982* : 71–93.

22. Here "consciousness" comprises Freud's "preconscious" system as well as his "conscious" system.

23. Here "the unconscious" comprises the Jungian collective unconscious and personal unconscious.

24. See note 10 above.

25. "No relation to an object is possible except mastery over it": Hegel, "On Love," trans. Richard Kroner, in *Early Theological Writings,* trans. T. M. Knox (Chicago: University of Chicago Press, 1948), p. 308. Cf. also Hegel's famous discussion of desire, work, and the master–slave dialectic in the *Phenomenology* (§§174–96).

26. Hegel, "On Love," p. 305.

27. Ibid., p. 304.

28. On the heart as a seat of knowledge as well as feeling, see James Hillman, *The Thought of the Heart,* Eranos Lectures 2 (Dallas: Spring Publications, 1984).

29. Hegel, "On Love," p. 305.

NIETZSCHE'S MADNESS AS SOUL-MAKING
A View contra Jung's

ROSS WOODMAN
(London, Ontario)

> Thus speaks the red judge, "Why did this criminal murder? He wanted to rob?" But I say unto you: his soul wanted blood, not robbery; he thirsted after the bliss of the knife. His poor reason, however, did not comprehend this madness and persuaded him: "What matters blood?" it asked; "don't you want at least to commit a robbery with it? To take revenge?" And he listened to his poor reason: its speech lay upon him like lead; so he robbed when he murdered. He did not want to be ashamed of his madness.
>
> *Thus Spake Zarathustra*

I.

In 1889, Nietzsche lapsed into complete insanity and, after brief sojourns in Basel, Jena and Naumburg, was taken by his widowed sister to Weimar where she nursed him until his death in August 1900. Born in 1844, the son of a Lutheran pastor with at least three generations of respected divines on both his father's and mother's side, he planned to study theology and classics at the University of Bonn. Disgusted by what he called "the bigoted Catholic population" of Bonn and passionately drawn to the culture of ancient Greece, in his second term he gave up theological studies to devote all his energy to classics. What he was looking for in turning from theology to classics, though at the time he didn't know it, was a *temenos* in which he could locate his own genius, that it might

---

Professor of English at the University of Western Ontario, London, Ontario, Canada, Ross Woodman is the author of *The Apocalyptic Vision in the Poetry of Shelley* (Univ. of Toronto Press). He has written numerous articles both on the English Romantics and contemporary art, particularly Canadian art.

flower and bear fruit in himself. His genius would release his century from the decadence into which he believed it had collapsed as a result of the collapse of Christianity itself, one so absolute that Nietzsche would sum it up in his announcement that "God is dead."[1]

With this announcement he sealed his fate. In tearing himself so completely from his inherited Christian perception of self, he found in its place a being so utterly new he was almost persuaded its like had never existed before. Nietzsche prophetically images this new being as the madman already at work within Nietzsche's various inventions of himself that would come together in the figure of Zarathustra, whose total derangement constitutes the chaos or void from which a new dancing humanity, Nietzsche's *übermensch,* radically discontinuous with its past, would emerge. "Whither is God?" cried the Madman in *The Gay Science.* Nietzsche's reply reveals the inherent madness in his announcement of the death of God.

> *We have killed him*—you and I. All of us are his murderers. But how have we done this? How were we able to drink up the sea? Who gave us the sponge to wipe away the entire horizon? What did we do when we unchained the earth from its sun? Whither is it moving now? Whither are we moving now? Away from all suns? Are we not plunging continually? Backward, sideward, forward, in all directions? Is there any up or down left? Are we not straying as through an infinite nothing? . . . Do we not smell anything yet of God's decomposition? . . . God is dead. God remains dead. And we have killed him. . . . There has never been a greater deed; and whoever will be born after us—for the sake of this deed he will be part of a higher history than all history hitherto. (p. 97)

This "higher history than all history hitherto" is the one man invents for himself which in the second edition of *The Gay Science* (1887) Nietzsche calls "a regulative fiction."[2] The new man, the *übermensch,* will locate himself not in a realm to which he attaches conviction or belief, but in a radical, provisional one in which for the moment he willingly and knowingly suspends his disbelief. Coleridge, the inventor of that phrase, distinguishes in *Biographia Literaria* (1817) between poetic faith and religious faith.[3] In his *Rime of the Ancient Mariner,* Coleridge persuades his reader to enter the world of his madman, not by satanically deceiving him with a false display of moral and religious truth, but by confronting him with what he calls "dramatic truth." The "excellence" of his poem,

Coleridge writes, "was to consist in the interesting of the [reader's] affections by the dramatic truth of [the mariner's] emotions [which] would naturally accompany such situations, supposing them real" (II, p. 6). What Coleridge as poet supposes to be real Coleridge as Christian knows or believes not to be.

Nietzsche rejects this Romantic distinction between poet and man. With the death of God, he argues, man must invent himself, inhabit the drama he himself writes as a "regulative fiction" sufficient to persuade him for the moment to suspend his disbelief. Madness for Nietzsche is the chaos or void, the mindless whirl of atoms on which a fictive order is momentarily imposed. This order constitutes for Nietzsche the relocation of man in the intermediate realm (neither matter nor spirit) of the psyche. It is within the precarious realm of what Keats called "Soul-making" (for Keats indistinguishable from poetry-making) that Nietzsche stakes his claim as "the first psychologist of Christianity."[4] By exploring in depth the "dramatic" (as opposed to religious) truth of the Christian myth, he restores it to its primitive condition of madness and to that "regulative fiction" it had temporarily imposed on the human soul.

The way in which that fiction hardened into dogma, he argues in *Thus Spake Zarathustra,* reduced the soul to the condition of a camel in the desert kneeling down to receive its burden (Zarathustra, the Greek form of Zoroaster, means "old camel"). In *Thus Spake Zarathustra* Nietzsche goes on to dramatize or enact the metamorphosis of the camel into a lion, a metamorphosis in which the camel is released from its burden and set free. Before this, however, the camel's burden must itself become a scaly serpent which the lion has to slay. To the dragon's "Thou shalt" the lion replies, "I will" (p. 139). The camel "once loved 'thou shalt' as most sacred: now he must find illusion and caprice even in the most sacred" (p. 139).

Once man is able to substitute "illusion and caprice" for what he formerly considered "the most sacred," he becomes a little child described by Zarathustra as a "forgetting, a new beginning, a game, a self-propelled wheel, a first movement" (p. 134). He becomes, that is, the *puer aeternus* who invents the game of creation, invents himself, invents the world. "Some souls one will never discover," Zarathustra explains, "unless one invents them first" (p. 154). The soul makes for itself a home in metaphor which both matter and spirit would turn into a dungeon by rendering metaphor literal. By Nietzsche's releasing soul from its warring enemies and relocating man in the fictions by which he lives, one might argue that he gave birth to the soul as an object of knowledge, gave birth, that is, to psychology.

II.

While Jung's indebtedness to Nietzsche cannot be fully appreciated until his lectures on *Thus Spake Zarathustra* are published, what Jung called Nietzsche's "exaggeration and irreality" filled him with horror. "Nietzsche," he remarks in his memoirs,

> had lost the ground under his feet because he possessed nothing more than the inner world of his thoughts—which incidentally possessed him more than he it. He was uprooted and hovered above the earth, and therefore he succumbed to exaggeration and irreality. For me such irreality was the quintessence of horror, for I aimed, after all, at *this* world and *this* life. No matter how deeply absorbed or how blown about I was, I always knew that everything I was experiencing was ultimately directed at this real life of mine.[5]

Jung makes it clear in his memoirs that, having made (as in Nietzsche's *Zarathustra*) what he calls "a voyage of discovery to the other pole of the world," he found he needed "a point of support in 'this world' " (p. 189). That support he found in his family, his medical diploma from a Swiss university, his patients, his home at 228 Seestrasse in Küsnacht with the servants and gardeners and oak paneling carved to his specifications, what he calls his "normal life in the real world," the "base to which he could always return" when "the unconscious contents could have driven [him] out of [his] wits." Nietzsche lacked this. Its absence drove him mad, that is, deeper into the unconscious beyond the point of return, there being no point to return to. He lacked a necessary material existence.

Was this the case? Did Nietzsche collapse into insanity having no way back or collapse into it because he had exhausted its creative possibilities, ending in his final ten years in a state of degenerative syphilis for which at the time there was no cure?[6] Was Jung's "normal life" the "real world"? Was Nietzsche's abnormal life "irreality"? Was the "quintessence of horror" for Jung his recognition of the essence of the autonomous soul? Did, for Jung, the essence of soul reside in Conrad's "heart of darkness" before which Conrad's Kurtz exclaims, "the horror! the horror!''? Was it the revelation of that soul essence in *Zarathustra* that drove Jung back to his ''normal life in the real world''?

One of what now seem the many ironies about the founders of psychoanalysis and depth psychology is their horror of the psyche if it is allowed to have its own autonomous way. After his break with Freud, when Jung began to write down his fantasies, he realized very quickly that he was in touch with what he called ''a personality that is not me, but which is insisting on coming through to expression'' (p. 185). More than that, he knew that the personality was that of a woman, indeed one of his own patients, ''a talented psychopath'' (p. 185) who had become a living figure in his own mind. Precisely in these terms—the soul as a psychopath—would Jung henceforth approach and deal with the soul. Not only was it necessary to ''differentiate oneself'' from the soul, but also by means of that differentiation to strip the soul of its power for, as he points out in his memoirs, ''the insinuations of the anima, the mouthpiece of the unconscious, can utterly destroy a man. . . . In the final analysis,'' he concludes, ''the decisive factor is always consciousness, which can understand the manifestations of the unconscious and take up a position toward them'' (p. 187).

That position was always to be on guard, lest as a psychopath the soul draw one into her own madness as, for Jung, she drew Nietzsche. Nothing better reveals Jung's ultimate fear of and contempt for the autonomous soul than his recorded rejection of it. ''Today,'' he writes, ''I no longer need these conversations with the anima, for I no longer have such emotions'' (p. 188). He would have the reader believe he had transcended or outgrown the soul to enter the absolute reality of pure spirit which, uncontaminated by the disease which is soul, Jung calls ''cosmogonic 'love' '' (p. 354). He put the word ''love'' in quotation marks to indicate that he did not use it ''in its connotations of desiring, preferring, favoring, wishing, and similar feelings, but as something superior to the individual, a unified and undivided whole'' (p. 354). It was against precisely that kind of transcendence that Nietzsche took his stand.

III.

The *deus absconditus,* the dark Dionysus who is hidden even from himself under the cloak of his own darkness, is everything which appears mindless, causeless, without precedent, unexpected. He induces shock, trauma, blindness, paralysis, migraine, stroke, seizure, allergy, schizophrenia, death. He is also the great awakener, shaking the mind loose from the lethargy of its armor, its security against the God's invasion. In sleep he speaks using words he does not understand (the God as God understands nothing). Nor does the dreamer understand. He listens, but he does not understand. He forgets. He awakens to a blank. "Your apparel, the armor that you walk in," speaks the voice from nowhere, "is the chains you wear. Arming yourself against me, you become me. You wear my darkness. You hide in my hiddenness. My chains declare you as the number in the German camp declares the Jew. You wear me as your skin. I become your skin. You long to shed your skin. You do. I'm the blankness when you awake, the sudden switch, your own perpetual absence."

Jung followed Nietzsche in search of the dark God. Nietzsche by way of Zarathustra embraced him, became his prophet. He baptized evil as his own highest good. Killing the Christian God brought his soul to life. Jung observed the death of God in the slow, painful death of his pastor father. His mother watched with him, his accomplice in his unconscious "Freudian" crime that would make him Freud's disciple. The crime—the wished-for death of the father (God shitting on Basel cathedral)—released him. "I gathered all my courage, as though I were about to leap forthwith into hell-fire, and let the thought come," writes Jung in his memoirs, the thought being that God wanted him to sin, that God spoke in the Devil, not to destroy him, but to bring him to a larger consciousness both of himself and of God.

"I saw before me the cathedral, the blue sky," he continues. "God sits on His golden throne, high above the world—and from under the throne an enormous turd falls upon the sparkling new roof, shatters it, and breaks the walls of the cathedral asunder." The result, however, was not Nietzsche's death of God, not the release that came for Nietzsche in the embrace of evil, an embrace that gave birth to the new man, Nietzsche's neopagan Heraclitean superman dancing in the eternal flux. It was, on the contrary, a new and fuller sense of God, a new birth of God within himself that included rather than excluded Satan, his elder son. It was, in short, a vision of God imaged as the reconciliation of the warring brothers, Cain and Abel, Jacob and Esau, Satan and Christ, the No. 1 and No. 2 personalities in Jung himself. It was a vision of wholeness described and experienced as grace. "So that was it!" Jung writes.

> I felt an enormous, an indescribable relief. Instead of the expected damnation, grace had come upon me, and with it an unutterable bliss such as I had never known. I wept for happiness and gratitude. The wisdom and goodness of God had been revealed to me now that I had yielded to His inexorable command. It was as though I had experienced an illumination. A great many things I had not previously understood became clear to me. That was what my father had not understood, I thought; he had failed to experience the will of God, had opposed it for the best reasons and out of the deepest faith. And that was why he had never experienced the miracle of grace which heals all and makes all comprehensible. He had taken the Bible's commandments as his guide; he believed in God as the Bible prescribed and as his forefathers had taught him. But he did not know the immediate living God who stands, omnipotent and free, above His Bible and His Church, who calls upon man to partake of His freedom, and can force him to renounce his own views and convictions in order to fulfill without reserve the command of God. In His trial of human courage God refuses to abide by traditions, no matter how sacred. In His omnipotence He will see to it that nothing really evil comes of such tests of courage. If one fulfills the will of God one can be sure of going the right way. (pp. 39–40)

Whatever momentary doubts Jung might have had throughout his life, particularly after his break with Freud, he was never very far from the conviction that whatever he did partook of divine freedom by submitting with courage to an omnipotent will. He remained persuaded that this will was itself "a *kosmogonos,* a creator and father–mother of all higher consciousness." Jung concludes his memoirs: "Whatever the learned interpretation may be of the sentence 'God is love,' the words affirm the *complexio oppositorum* of the Godhead" (p. 353).

Though Jung made much of the *deus absconditus,* particularly in his *Answer to Job,* he like Job was never persuaded that God's darkness was anything more than his omnipotent light shining in a world that could not comprehend it. To entertain even the possibility that the darkness might be other than blindness, blind unconsciousness, was to invite the madness of Nietzsche. Such a thought or realization would constitute "the quintessence of horror."

Jung's quarrel with Father White over the Catholic view of evil as the *privatio boni* (evil as the absence of good rather than an autonomous force independent of it) was, as Father White always suspected, specious, not really about the nature of evil at all, for both affirmed the omnipotence of love. It was, on the contrary, Jung's deep-rooted Protestantism rebelling against Catholic dogma that placed institutionalized experience of the church above the individual's experience of the will of God. It was, in short, his quarrel with his father. He opposed Father White's orthodoxy on the basis of his own experience without realizing that it finally supported that orthodoxy. To realize this would have been to render his own experience answerable to the church, to look beyond himself for a validation of himself, which would in turn rob him of his passion, his own unique relationship to God. It would be to share his beloved mother with another, something that Jung in his life never had to do and would not. Father White confronted in Jung an impregnable ego, what Keats—referring to Wordsworth—called "the egotistical sublime."[7] It was the invincible Protestant spirit that Jung, confronted by his father's spiritual exhaustion, was determined to renew. Unlike Nietzsche and in the spirit of Calvin and Luther, Jung remained firmly in the Protestant Christian fold.

IV.

The problematic nature of the soul lies in the fact that it has always been considered an appearance rather than a reality. The soul images or mirrors, reflecting a reality not its own, one that does not take up residence in it. Nothing inhabits the soul, not even the soul itself. There is nothing behind the mirror or in the mirror. The belief that there is (Alice's experience in Wonderland) is a form of lunacy, the confusion of an image for the reality, identification with the image as the reality, the mirror as narcissistic possession. Hillman begins the first part of "Silver and the White Earth" by quoting what Hegel is supposed to have said in a lecture: " 'in insanity the soul strives to restore itself to perfect inner harmony.' "[8] "For Hegel," Hillman explains, "insanity is a necessary stage or form of the soul, and a stage upon which the soul must perform. Insanity is essential to soul-making" (p. 21).

Nietzsche's insanity, which Jung found the "quintessence of horror," was a stage erected for the soul's performance, the Dionysian theater in which his soul enacted itself, much as the soul had enacted itself in the theater of Aeschylus. Nietzsche realized he had to erect this Hellenic stage because the soul could not perform, enact itself, on the Christian altar. On this altar the soul is murdered. The mass no less than Nietzsche declared that God is dead. What killed him was what killed the soul: the incarnation in matter. The world as matter murdered the soul even as the world as spirit murdered the soul. Between matter and spirit, the soul was trapped in a double bind. To affirm the one or the other was to die. In that double bind of matter and spirit the soul has no alternative, if it is to live, but to go mad. "In insanity the soul strives to restore itself to perfect inner harmony." Nietzsche's birth of soul, his soul-making, is his own birth of tragedy enacted upon a stage where his madness was free to perform. That tragedy, a rebirth or recurrence of Greek tragedy, was a vision of insanity restored to "perfect inner harmony," a coming together of Dionysus and Apollo.

Thus Nietzsche rejected the conventional view, as developed by Winkelmann and Goethe, of the ancient Greeks as harmonious. On the contrary, Nietzsche argued (taking his cue from Burkhardt), the Greeks acted out of an envy, jealousy, hatred and ambition that brought them again and again to their own destruction through deeds of *hybris.* In that *hybris* lay a madness embodied in the Furies that whirled them about like atoms in the void even as it whirled Orestes after he slew his mother at Apollo's command. In Aeschylus this madness was restored to "perfect inner harmony" (Orestes is acquitted by the Areopagus) beyond pessimism and optimism, beyond the blinding, migrained experience that on one level was Nietzsche's life. Dionysian frenzy or madness raised by Apollo to harmony through a sanctioning recognition of its necessity was what Nietzsche found in Greek tragedy and hoped to achieve in himself: to embrace the frenzy rather than reject it, to let the harmony rise in it, spring out of it, to make the two Gods one. The birth of tragedy from the spirit of music. Zarathustra dancing.

V.

Nietzsche knew his alchemy. All heterodox visionaries do. The Incarnation was the *nigredo,* God in the lead mine, dead. Nietzsche constructed himself, made for himself a subtle body, by mining the lead for silver, mining the dead God, the carcass of a decaying God. Out of God's putrefaction in his own stomach and brain he made silver. He made soul.

The lead mine is nature itself, the world as body, as flesh. "Antiquity, that is Lucretius, Strabo, Diodorus," writes Hillman, "considered silver to result from a forest fire, a holocaust. A gigantic fire rages through a forest, charring the greenwood, decimating nature, and after that ruin a thin stream of silver emerges. Silver, they infer, comes out of gigantic psychic disasters. It results from burnout. We claim it only after the woody bowers of protective naturalism have been totally blackened'' (pp. 32–33). The burnout to which he refers—the blackened nature, the blackened Madonna—points to the furnaces in which the silver is extracted from smelted lead. It is an art that must be learned. What is required is repeated calcination (drying heat), quenching the hot amalgam of silver and lead ore in a sharp liquid (sal ammoniac or vinegar), and distillation requiring calcination again. It is not easy. For a few ounces of silver, a ton of smelted lead is usually required. That lead remains as heavy depression. ''This is surely familiar,'' writes Hillman.

> An insight may be shining in itself, but it makes no dent on the gray mood from which it came. We have the feeling of lead-poisoning, a state of being swallowed by lead, lost in the lead mine—it's all so long and slow. What use, we feel, these silver mirrorings, this trickle of light, if the caughtness of depression does not lift. The truth, however, is that *the depression is the mine.* This is the lead necessary for the silver, even if the silver is neither useful nor functional in altering the lead. We mine silver from lead, but not to do away with the lead for that would close down the mine. (p. 34)

The dead God was Nietzsche's lead mine. Nothing could alter that death, for to alter it would do away with the mine. The dead God was Nietzsche's ore. It was also his crucified body. Nietzsche mined his own sickness, mined his migraines for silver as he mined his phlegm. ''The perfect lightness and levity, even exuberance of the spirit,'' he writes in *Ecce Homo,* ''are quite compatible in my case not only with the deepest physiological weakness, but even with excessive pain. Amid the tortures that go with an uninterrupted three-day migraine and agonizing phlegm-retching, I possessed a dialectician's clarity par excellence, and very cold-bloodedly thought through matters for which, in healthier states, I am not enough of a climber, not subtle, not *cold* enough'' (pp. 657–58). At last, he had got all the silver there was and died, as Zarathustra advised, at the right time. He had what Nietzsche called a free death, the death he desired because it consummated his life. Nietzsche had from the start of his philosophical career ''thirsted for the bliss of the knife.'' Though at times his ''poor reason, whose speech lay upon him like lead'' told him he wanted to commit robbery, take revenge, it was murder he was after, and got.

The alchemists imaged the release of silver from lead as a white dove (a leaden state is a dove solidified, a bird without wings, a gull in an oil spill). Nietzsche's Zarathustra emerges at the end of the fourth book from his mine-cave to hear himself "surrounded by innumerable swarming and fluttering birds." These fluttering birds about his head are his wisdom which he had used to kill the enemy, to kill Christianity. Thus he compares them to "a cloud of arrows that empties itself over a new enemy." Now, they become "a cloud of love, and over a new friend" (p. 450). Assigning a function to hatred (which Nietzsche considered essential to soul-making), Hillman sums up much of Nietzsche:

> Perhaps our hatreds are not personal only, but necessities of silver, firming the mind with supportive principles, giving determinate backing to its mirror. For in love we may not see; in hatred every line of the other stands out in cruel detail. Maybe hating belongs to silver-making, and is a requirement for the conjunction with gold, where the vision of perfection gilds all things with a golden glow. Maybe the conjunction also means a conjunction of the *shadows* of gold and silver—golden love tempered by a hatred which cools the heart's optimism with astringent perceptions like a cold silvered glass at the back of the mind. The work with silver polishes the hatred. The personal salty nature of hating and its sulphuric compulsion become "translucent," so that the personal object of hatred and the personal feelings resolve their narrowness of focus into the face of an image, like a frightful Japanese, Tibetan or Greek mask. Not harden the heart, but harden the mind, for the courage and faith which gold brings assumes true value only when it is married with these cold depths of earthed insight that hatred so well provides. (p. 39)

Jung did not mine a dead God; he mined a sleeping one by analyzing his dreams. The Christian God became Jung's analysand. And he brought him into a new consciousness of himself, helped him to embrace his shadow as his friend. Unlike Nietzsche, Jung was committed finally not to the soul but to the spirit. Nietzsche used spirit as he used matter: he mined it for soul. What Nietzsche was after was not gold but silver. Jung was after the gold. His real object was God.

Not only did Nietzsche take up residence in soul, he released it as a phenomenological entity, as the world in which all things appear to become incarnate not in flesh but in metaphor. Metaphor—the subtle body, soul—is the mirror in which all things appear, the mode of all human perception. We live, Nietzsche concludes, in images which are all things beheld and beholden. Only in and by the conscious process of beholding, which Coleridge calls the "primary imagination" or "living power and prime agent of all human perception" (I, p. 304), does experience take place; only in and through experience does the world for man reside. All else, whether matter or spirit, is death. That is why, for Nietzsche, the incarnate God, which is matter as spirit or spirit as matter, is dead.

VI.

"Here," writes Henry Corbin at the conclusion of his essay "Divine Epiphany and Spiritual Birth,"

> it is not our task to apprehend in terms of pure phenomenology the consequences, *a parte Dei,* of homoousia and incarnation. God has ceased to be the Eternal in Heaven. Incarnated, He is buried in the flesh until the time comes for Him to be buried in the grave. And this Incarnation was the full Revelation of the godhead. Hence there is no more mystery; the mystery has been proclaimed on the housetops, it has been "profaned." Still, in terms of pure phenomenology this signifies identification to the point of perfect coincidence between consciousness and transconsciousness, a coincidence that was to have incommensurable consequences for the spiritual history of our Occident.[9]

One of those consequences was a radical materialism, an empiricism of the flesh which identified God with matter (buried in matter as in a grave), and then turned matter itself into God. God, by entering history in and as the body of Jesus, *became* history even as he entered matter to become matter. This radical presence—homoousia and incarnation—of the body of God in and as history, a presence affirmed and perpetuated on the Catholic altar through eucharistic transubstantiation, became with the death of God Marx's dialectical materialism, a heresy the church fathers gathered in council could no longer control.

Nietzsche rejects the Incarnation not only in the person of Christ but also in history. For Nietzsche not only God was dead; history had died with him. "His two books *The Anti-Christ* and *Ecce Homo,*" writes Janko Lavrin, "are violent attacks not only on Christianity, the 'one great curse of mankind,' but—implicitly—on Christ and Christ's place in history. What Nietzsche now aimed at was to achieve as great a revolution in the history of mankind as the one achieved by Christ, but in the opposite direction."[10] Nietzsche believed that in his work, which by 1888 he was in a position to sum up, he had cleaved the history of mankind in two even as Christ had. Formerly history was divided into B.C. and A.D., henceforth B.N. and A.N. (before and after Nietzsche). Speaking of his *Transvaluation of Values,* addressed to what he calls "the very few" (none of whom may be living yet) and left largely unwritten (though it lay ready before him), he swore that "in two years time the whole of our earth will be in convulsion." "I am," he cried, "fatality itself." "I literally have to bear the fate of all mankind," he wrote to the editor of the periodical *Kunstwart.* "In two months I shall be the foremost man on earth," he told Professor Overbeck toward the end of 1888. On December 31 of that year, he sent August Strindberg a message: "I have," he wrote, "appointed a meeting-day for the monarchs in Rome. I will order the young Kaiser to be shot."[11] Here, if anywhere, lay for Jung "the quintessence of [Nietzsche's] horror."

Nietzsche's "insane" pronouncements, however, were no more insane than those of Hegel who also declared that history had ended, the world soul having restored itself "to perfect inner harmony" in the Prussian state after the long period of "insanity" that constituted all that preceded it, a history Joyce would describe as a nightmare from which we are struggling to awake. What, that is, is the "convulsion" of "the whole of our earth" that Nietzsche was persuaded he was enacting in his work? What is at work in Marx's vision of dialectical materialism that declares the end of history in a stateless society governed by a self-sustaining inner dynamic as harmonious in its operations as Hegel's world-spirit? What led the Persian prophet, Baha'u'llah, arising in 1853 out of Shiite Islam as the promised Imam Mahdi, the Resurrector, to declare, "Lo, the entire creation hath passed away! Nothing remaineth except my Face..."?[12] What, it must be asked, is going on here if not "a quintessence of horror"? What in short is the insanity of the nineteenth century enacting?

I believe Nietzsche provides the most plausible answer. "Well," he wrote to Brandes in November 1888, "I am the first psychologist of Christianity, and as an old artillery man, I can turn on it my heavy guns, the existence of which has not even been suspected by the enemies of Christianity." What the nineteenth century was enacting was the psyche. It was relocating man and history in the psyche itself rather than in spirit or matter. The nineteenth century was the first century in world history committed to soul-making, which is to say, to an understanding of the world as soul, as a subtle or soul-body rather than physical body, as imaginal, as dream understood as the reality of the imaginal, the reality of dream. The nineteenth century was man's awakening (in figures like Hegel and Nietzsche) to the imaginal to find it true. In the insanity of Nietzsche, the nineteenth century was giving birth to psychology, giving birth to psychic reality as the phenomenology not of the spirit but of the soul, not of God (revelation) but of man. When Nietzsche says he is "the first psychologist of Christianity," he means that he mined the Incarnation and found the soul buried there, as silver is buried in lead. As what in *Nietzsche Contra Wagner* he calls the "unriddler of souls" (p. 677), he entered the white earth.

To argue that Nietzsche in mining the dead God (which is what he took the Incarnation to be) was releasing the soul is to argue that he saw himself as the Resurrector. What, however, he was resurrecting was not God but humanity, though to be sure a radically new human being who fully inhabited not spirit, which is the death of the human, but soul which is his resurrected life. "But now this god has died," proclaimed Zarathustra to those whom he has gathered in his cave.

> You higher men, this god was your greatest danger. It is only since he lies in his tomb that you have been resurrected. Only now the great noon comes; only now the higher man becomes—lord.
>
> Have you understood this word, O my brothers? You are startled? Do your hearts become giddy? Does the abyss yawn before you? Does the hellhound howl at you? Well then, you higher men! Only now is the mountain of man's future in labor. God died: now we want the overman [*übermensch*] to live. (pp. 398–99)

Nietzsche's *übermensch* is the soul enacting in a theater of madness its own birth, birthright, rite of birth.[13]

VII.

For the extinct fiat "Be," Nietzsche substituted "It is"; for the extinct fiat "Thou shalt," he substituted "I will." This radical substitution of the human for the divine arises in Nietzsche's philosophy from his conviction that God is dead. In the place of "Be" and "Thou shalt" stood the whirl of atoms in the void, itself a metaphor of madness where the hammer falls from the philosopher's hands.[14] Human action is robbed of any metaphysical sanction. Man has to invent himself, there being no other inventor.

Mythologies are counters to play with, and in the spirit of play (Beckett's "Me to play"[15]) Nietzsche, like Blake before him, exploits the radical Christian opposition between God and Satan, good and evil, love and hate, attraction and repulsion, not in order to eradicate the one in favor of the other but to liberate the psychic dynamics of an ascending life. "Without Contraries is no progression," writes Blake in *The Marriage of*

Heaven and Hell;[16] "Attraction and Repulsion, Reason and Energy, Love and Hate, are necessary to Human existence." By rejecting this necessity through its radical polarization of the "Contraries" into good and evil, Christianity (Blake argued, as did Jung after him) rejected "Human existence" itself, an existence in his view dependent on the lively, continuous and creative interaction of the "Contraries." The tyranny of the "Good" lay in its rejection of "Evil," by promulgating the belief "that God will torment Man in Eternity for following his Energies." Blake accordingly embraced the "torment" even as Nietzsche embraced his body, declaring that "torment" to be his own "Energies." "Energy," Blake declares, "is the only life and is from the Body. . . . Energy is Eternal Delight."

Torment as "Eternal Delight" characterizes the work of both Blake and Nietzsche. Walking among the "fires of Hell," Blake is "delighted with the enjoyments of Genius; which to Angels look like torment and insanity." Working in his "Printing House in Hell," he describes his illuminated works as "printing in the infernal method, by corrosives, which in Hell are salutary and medicinal, melting apparent surfaces away and displaying the infinite which was hid." Describing to an Angel how Jesus Christ murdered the Jewish God by breaking all his commandments and living by "impulse" rather than "rules," Blake, speaking as "the voice of the Devil," converts the Angel to the Devil's party. "This Angel, who is now become a Devil," he writes, "is my particular friend: we often read the Bible together in its infernal or diabolical sense which the world shall have if they behave well." This Blakean Bible is a series of "poetic tales" constructed by the human imagination which is the proper and only dwelling place of God. To remove God from the imagination by turning him over to a priesthood whose function is to enact daily upon the altar its murder of him is to invent religion. Christianity thus becomes for Blake a priestly system which perverted "poetic tales" into "forms of worship" by removing God from the imagination much as a fish is removed from water so as to kill and eat it. "Til a system was formed," writes Blake, "which some took advantage of & enslav'd the vulgar by attempting to realize or abstract the mental deities from their objects: thus began Priesthood."

Blake's task as a poet belonging to the Devil's party is to restore God to life by removing him from the sepulcher of the priestly altar and implanting him again in the human imagination where he can truly incarnate himself in works of art, in the construction of a body of metaphor, a soul or "subtle" body which is the body of the imagination. Blake locates God in the soul-making process, a process which requires for its life a ceaseless war of "Contraries" such as he, and after him Nietzsche, enacts. That war, properly understood, transforms Christian "torment" into the "eternal delight" celebrated by the dancing Zarathustra who turns Christ's last Supper into a Dionysian feast. Breaking free of "forms of worship," treating priesthood in a scatological manner (finding soul in God's excrement), Blake and Nietzsche celebrated the resurrected life of man arising out of the death of God as the Christians had for millennia sacramentally performed it. That celebration—the ultimate necessity of the "Contraries" and the consequent loving embrace of them—Nietzsche called *amor fati*. "My formula for greatness in a human being is *amor fati*," he writes in *Ecce Homo*, "that one wants nothing to be different, not forward, not backward, not in all eternity. Not merely bear what is necessary, still less conceal it—all idealism is mendaciousness in the face of what is necessity—but love it."[17]

Nietzsche's madness must finally be understood as Dionysian abundance, a soul-life experienced as an overflowing cup in which as in a chalice torment became delight. "You higher men, what do you think?" asks Zarathustra.

> Am I a soothsayer? A dreamer? A drunkard? An Interpreter of dreams? A midnight bell? A drop of dew? A haze and fragrance of eternity? Do you not hear it? Do you not smell it? Just now my world became perfect; midnight too is noon; pain too is joy; curses too are a blessing; night too is a sun—go away or you will learn: a sage too is a fool.
>
> Have you ever said Yes to a single joy? O my friends, then you said Yes too to *all* woe. All things are entangled, ensnared, enamored; if ever you said, "You please me, happiness! Abide, moment!" then you wanted *all* back. All anew, all eternally, all entangled, ensnared, enamored—oh, then you *loved* the world. Eternal ones, love it eternally and evermore; and to woe too, you say: go, but return! *For all joy wants—eternity.* (p. 435)

The eternity that joy wants it receives because joy wants it. The reception resides in the desire itself, desire creating for Nietzsche its own object which, while sensate like the objects in a dream, is not perceptual; it does not exist out there. It exists in the desire itself as its imaginal body. The recognition of that body as the soul's body rather than its borrowed natural apparel constitutes the soul's unveiling to reveal not nothingness but nakedness, in which resides for Nietzsche the soul's "eternity." Wrapped in that eternity or nakedness containing within itself the principle of its own activity, the soul rejoices in its own body of metaphor that, free at last of the literal, becomes its *opus contra naturam,* the pains and pleasures of which are as remote from the literal as metaphor from the actual. Lear's agony on the heath is an act of joy. To say "Yes" to that *opus contra naturam* is to say "Yes too to *all* woe." Mad Lear naked on the heath is for Nietzsche the soul dancing.

1. "The greatest recent event—that 'God is dead,' that the belief in the Christian God has ceased to be believable—," Nietzsche writes in *The Gay Science,* "is even now beginning to cast its first shadows over Europe. For the few, at least, whose eyes, whose *suspicion* in their eyes, is strong and sensitive enough for the spectacle, some sun seems to have set just now. . . . in the main, however, this may be said: that the event itself is much too great, too distant, too far from the comprehension of the many even for the tidings of it to be thought of as having *arrived* yet, not to speak of the notion that many people might know what has really happened here, and what must collapse now that this belief has been undermined—all that was built upon, grew into it; for example, our whole European morality. . ." (p. 441). All quotations from Nietzsche, unless otherwise noted, are taken from *The Portable Nietzsche,* ed. and trans. Walter Kaufmann (New York: The Viking Press, 1968).

2. Nietzsche is conscious of the fact that the new or "gay" science still takes its "flame" from a fire "which a faith thousands of years old has kindled: the Christian faith, which was also Plato's faith, that God is truth, that truth is divine. . ." (p. 450). It still stands "on the ground of morality," a morality that constitutes a "will to truth," a "refusal to be deceived" (p. 449). That ground, he insists, must be removed so that "error, simulation, deception, self-deception" (in a word, fiction) may become the proper home of the soul which, liberated from both matter and spirit, must now be built by the "godless ones and anti-metaphysicians" (p. 450). Fiction as the truth of the soul (psychic reality) must replace metaphysics.

3. *Biographia Literaria,* ed. James Engell and W. Jackson Bate, 2 vols. (Princeton: Princeton University Press, 1983), 2 : 6.

4. Letter to Georges Brandes, November 1888. Quoted in Janko Lavrin, *Nietzsche: A Biographical Introduction* (London: Studio Vista, 1971), p. 110.

5. *Memories, Dreams, Reflections,* ed. Aniela Jaffé (New York: Pantheon Books, 1963), p. 189.

6. "The entry 'syphlit. infection' has raised rather more dust than was necessary," writes E. F. Podach in *The Madness of Nietzsche,* trans. F. A. Voight (New York: Gordon Press, 1974). "There is no certain evidence of such an infection. Nor is there any sure evidence of Nietzsche's 'paralysis.' In all probability there can be no definite verdict on the matter because of the inadequacy of the data" (p. 235). Podach's treatment of Nietzsche's madness supports in part the view taken in this paper. "In considering Nietzsche even in his illness the author would appeal to those who, in the words of Novalis, hold that 'the life of a truly canonical man must be symbolical from beginning to end,' " Podach writes in his preface to the English edition (p. 16).

7. Letter to Richard Woodhouse (27 October 1818) in *The Letters of John Keats,* ed. H. E. Rollins, vol. 1 (Cambridge: Harvard University Press, 1958), p. 387.

8. *Spring 1980* : 2. All quotations from Hillman, unless otherwise noted, are taken from this essay.

9. "Divine Epiphany and Spiritual Birth in Ismailian Gnosis," in *Man and Transformation,* ed. Joseph Campbell (Princeton: Princeton University Press, 1972), pp. 58–59.

10. Lavrin, *Nietzsche,* pp. 109-10.

11. Ibid., pp. 110–11.

12. *Gleanings from the Writings of Baha'u'llah* (Baha'i Publishing Trust, 1949), p. 29.

13. Soul-making for Nietzsche is the making of the *übermensch,* the new man as opposed to what he calls the historical man, the man who is the slave of historical culture. Nietzsche in his preface to *The Use and Abuse of History* begins by quoting Goethe: "'I hate everything that merely instructs me without increasing or directly quickening my activity.'" He then adds by way of comment: "I will show in them [Goethe's words] why instruction that does not 'quicken,' knowledge that slackens the reign of activity, why in fact history, in Goethe's phrase, must be seriously 'hated,' as a costly and superfluous luxury of the understanding: for we are still in want of the necessaries of life, and the superfluous is an enemy of the necessary. . . . In other words, we need [history] for life and action, not as a convenient way to avoid life and action, or to excuse a selfish life and a cowardly base of action. We would serve history only as far as it serves life; but to value its study beyond a certain point mutilates and degrades life." Nietzsche reads history allegorically as soul-events taking place in great men. Unless it quickens soul-events, it kills the soul by overwhelming it with a panorama of events that turns it into "a restless, dilettante spectator" where even "great wars and revolutions cannot affect [it] beyond the moment." "The war," he writes, "is hardly at an end, and it is already converted into thousands of copies of printed matter, and will be soon served up as the latest means of tickling the jaded palates of the historical gourmets" (p. 39). It is precisely this passive treatment of history as the culture of jaded palates that Nietzsche opposes in favor of soul-making.

14. Nietzsche's alternative title for *Twilight of the Idols* was *How One Philosophizes with a Hammer.* I am suggesting that the void that replaces God is the insanity that the idea of God contains or controls. Nietzsche describes the death of God as insanity. The murderer of God is a madman who, not wanting to be ashamed of his madness, invents sane reasons for his murder, even as Christians invent sane reasons for God's murder of his Son. Nietzsche contained his madness not by the ritual games of Christian belief but by the ritual play of the imagination which constitutes art. In madness, Foucault argues in *Madness and Civilization* (New York: Vintage Books, 1973), art experiences its own absence, madness being "the absolute break with a work of art" (p. 387). What is absent in the work of art is the subject itself liberated from its *temenos* or container which Shelley, for example, considered "a feeble shadow of the original conception of the poet." Nietzsche's subject, present as shadow-boxing in his work, is the identity of Christ and Dionysus. Thus Foucault writes: "Nietzsche's last cry, proclaiming himself both Christ and Dionysus, is not on the border of reason and unreason, in the perspective of the work of art, their common dream, finally realized and immediately vanishing, of a reconciliation of the 'shepherds of Arcady and the fishermen of Tiberias'; it is the very annihilation of the work of art, the point where it becomes impossible and where it must fall silent. . ." (p. 287).

15. *Endgame* (New York: Grove Press, Inc., 1970), p. 68.

16. *The Complete Poetry and Prose of William Blake,* ed. David Erdman (Berkeley: University of California Press, 1982).

17. *Basic Writings of Nietzsche,* trans. Walter Kaufmann (New York: Random House, 1968), p. 714.

COMBAT NIGHTMARES
Toward a Therapy of Violence

HARRY A. WILMER
(Salado, Texas)

Like an eyeless nightmare,
Grief did sit upon his being.
Shelley, *Pr. Athan*

It is urgently important to examine ways in which the human mind processes images of catastrophic violence, torture, and terrorism. Imagine the effect of hitting a large granite stone with a hammer, and the stone breaking along its natural cleavage veins in accordance with the power and angle of the blow. This metaphor conveys the way in which mental images of catastrophic trauma are perceived by the survivors in their nightmares, flashbacks, and ruminative intrusive thoughts. It is not a bad image if you can imagine yourself an unsuspecting stone.

The characteristic traumatic nightmare of combat is the hallmark of posttraumatic stress disorder, in the past known as shell shock, battle fatigue, and war neurosis.[1] In these dreams the images are reported to replicate exactly the trauma and nothing else.

Such dreams are unique, predictable in content, recurrent, and highly disturbing. They may lead to explosive behavior. These dreams cannot be explained satisfactorily by any traditional dream theory. This phenomenon has not been studied in any systematic way dealing with the form and content of the dreams. There is, however, increasing appreciation of the im-

---

Dr. Wilmer presented versions of this lecture at the 53rd International Meeting of the Mayo Alumni Association at Rochester, Minnesota, on October 11, 1985, and at the Annual Meeting of the Texas Psychiatric Society at Corpus Christi, Texas, on November 10, 1985. He is a Jungian analyst in private practice; president, director and founder of the Institute for the Humanities at Salado; and professor of psychiatry at the University of Texas Health Science Center at San Antonio. He has published an article in *Spring 1977*.

portance of war dreams to understand posttraumatic stress disorder.[2] There are other war dreams in which the exact trauma appears in altered form as if the psyche were processing the mental images and transforming them into ordinary nightmares. Although mental images in this disorder have been generally overlooked,[3] I have published on psychological characteristics of posttraumatic stress disorder,[4] and on the description, management and meaning of Vietnam war dreams in individual and group psychotherapy.[5]

The present paper reports a study in classification of dream format and motif toward a therapy of violence. Jung wrote,

> ... as to the *classification of dreams,* I would not put too high a value on the practical or on the theoretical importance of this question. I investigate yearly some fifteen hundred to two thousand dreams, and on the basis of this experience I can assert that typical dreams do actually exist. But they are not very frequent, and from the final point of view they lose much of the importance which a causal standpoint attaches to them on account of the fixed significance of symbols. It seems to me that *typical motifs* in dreams are of such greater importance since they permit comparison with motifs of mythology.[6]

Neither Jung nor Freud studied war nightmares of combat veterans. The typical catastrophic nightmare of the trauma is frequent, and its form is memorable because it is unique in dream life. The mythological nature of such dreams is rich, but beyond this paper. The evidence in this study supports the view that these dreams are neglected, vital elements in the therapy of violence.

The Study and the Sample

I devoted one full sabbatical year entirely to interviewing and therapy of combat Vietnam veterans at the Audie Murphy Veterans Hospital in San Antonio, Texas. A second year was devoted to studying the transcripts of these sessions, reviewing the audiotapes and writing the observations. That this work was painful and stressful goes without saying.

The veterans were referred to me by psychiatric and non-psychiatric inpatient services, by various outpatient services, the VA Vets Center outreach program in downtown San Antonio, and by veterans whom I saw. I conducted all interviews myself. To those men in whom the interview stirred up symptoms and who did not otherwise have access to psychotherapy, I offered psychotherapy. The average interview lasted three hours over two or three days; therapy lasted from a few sessions to over two years. This is not a random sample, but consists of men who saw themselves as psychologically troubled and in need of help from a psychiatrist working at the VA on a Vietnam project. No women were referred to me. Each subject signed an official VA informed consent form, approving audiotaping for use in articles and books. Over 300 hours were transcribed, providing verbatim records which are cited in this paper. The study is, therefore, based on 103 combat veterans who told me 359 dreams of Vietnam.

It should be noted that this study was conducted seven to sixteen years after the veterans had returned to the United States from Vietnam. Twelve of the men did not recall any dreams about Vietnam. It is unlikely that they had not dreamed of Vietnam, but only the remembered dreams were noted and the veterans were not pressed for dreams. The ninety-one remaining veterans reported from one to sixteen individual war dreams. Recurrent identical dreams were counted only once although they had occurred hundreds or thousands of times. The average number of dreams per veteran was 3.48.

There was no systematic attempt to get a random sample, and any veteran who came to see me was interviewed. Twenty-six were not taken into the study because they had not seen combat in Vietnam or were acutely psychotic and unable to give a coherent interview. Two veterans refused to sign the informed consent, and two veterans were obviously seeking me only as a means to get compensation and their stories were unreliable. All others were accepted. My interviews and records were not a part of the official VA records, so they could in no way influence whether or not the veteran might receive compensation.

How well does this sample of patients represent the Vietnam veterans who were in Vietnam each year of the war? There is an astonishing correlation although the patient sample was overrepresentative of men in Vietnam at the time of the Tet offensive, the turning point of the war (Table I).

No VA hospital record of any one of these men contained any information on any nightmare, and none contained any report of their Vietnam military experience. So that I had not made a biased observation, my research assistant reviewed all of the records with the same finding. Almost all of these veterans said that their doctors had not asked them about their war dreams or war experiences, and they had not volunteered this information.

TABLE I.

Observed and Expected Proportion of Subjects in the Year They Served in Vietnam (N=96)

	OBS.	EXP.	X^2	Total U.S. Personnel in Vietnam as of Dec. 31*
1964	2	.86	1.50	23,300
1965	7	6.82	.00	184,300
1966	17	14.26	.53	385,300
1967	32	17.97	10.95	485,600
1968	21	19.84	.07	536,100
1969	18	17.59	.01	475,200
1970	13	12.38	.03	334,600
1971	5	5.80	.11	156,800
1972	1	.90	.01	24,200
1973	0	.01	.01	Less than 250

X^2 = 15.8887 (for both in and out) $X^2 = E \frac{(obs - exp)^2}{exp}$

OBSERVED = men in sample in Vietnam on December 31.
EXPECTED = proportion based on total number of men serving in Vietnam during war.

The overall X^2 does not depart significantly from chance—we may therefore conclude that the distribution of men in the sample does not deviate significantly from a random sample drawn from men who served in Vietnam.

The major deviation in these data is in the number of men in the sample who were in Vietnam 12/31/68. These would have been veterans who most likely were in combat during the time around the Tet offensive.

* Source: Department of Defense, OASD (Comptroller) Directorate for Information Operation, Selected Manpower Statistics.

One veteran told me, "I think the VA would like to take all of the Vietnam veterans and put us in a closet and just lock us up behind a door and forget about all of it." This reflected the attitude not only of the professionals but of the country itself. After the war, people wanted to forget and to look away from the tragedy of Vietnam. This collective attitude of denial gradually eroded, and in the years since the dedication of the Vietnam memorial in Washington, D.C. in 1982, there has been a growing positive feeling toward the Vietnam veterans, as well as a great number of books, films, and studies of the war and the veteran's problems.

One of the obvious reasons we didn't want to hear their dreams and their experiences was that we didn't want to hear the grisly facts of a war we lost. With psychotherapists a countertransference is evoked which one veteran expressed like this: "We want to let them know what actually happened and what caused the Vietnam veterans to be as they are. It's gonna be hard for them to understand how Americans could behave like savages when, of course, they would have behaved exactly the same way if they were over there. Yeah, that's the way they'd have behaved."

Demography

Two-thirds of these veterans volunteered for military service. They spent from 2 to 32 months in Vietnam with a mean time of 14.3 months. Their mean age was 33.9 years. They were 33.2 percent Hispanic, 52.7 percent Caucasian, 7.3 percent Black; others not known. The high percentage of Hispanics reflects the San Antonio population with 53 percent, and Texas 21 percent; the national figure was 4 percent. Distribution among branches of the service was: Army 67.3 percent; Marines 20.0 percent; Navy 10.9 percent; and Air Force 1.8 percent. With the exception of one officer, all were enlisted men. Pre-service attitudes toward war were: patriotic and committed, 70.9 percent; opposed to the war, 7.3 percent; and no opinion, 21.8 percent. The mean number of years spent in school was 12.7 years. Grade school drop-outs were 38.2 percent. Service-connected disabilities were 26.3 percent. Divorced one or more times, 29.1 percent.

The interviewing was done between 1980 and 1981. At the beginning of the study, the diagnosis of posttraumatic stress disorder (PTSD) was not officially recognized by the Veterans Administration and the diagnosis was rarely used at this hospital, so there is no reliable evidence of the diagnosed incidence of PTSD.

The Dream

Although the combat dreams constituted only a part of the long interview, the veterans knew of my particular interest in this area. After the patient told his dream and elaborated on it, he was invited to give his interpretation, his own impression of the dream's meaning. There was an explicit assumption that dreams have meaning that should be ultimately helpful, and that combat dreams in particular might illuminate the veteran's understanding of his war trauma. While I naturally held a basically Jungian attitude toward dream meaning, I kept as unbiased an attitude as possible by adhering to the method of multiple working hypotheses.[7] We are dealing with vivid images, powerful narrative, and tremendous affect, so I relied heavily on the veteran's ideas about his dream and his metaphors and rarely indulged in deep interpretations.

The sample of 359 dreams represents as well thousands of recurrent dreams. Although the veterans were exposed to multiple war traumata, usually one event was so deeply engrained in memory that it alone constituted the basis of their characteristic dream, expressing how it actually had happened with attendant feelings. There is no conceivable way of checking exactly how the dream images followed the event. In rare instances where it seemed obvious that the dream was embellished or partly fabricated, such dreams and "war stories" were not included in the study. With patients who were psychotic, there was little difficulty in distinguishing a natural dream from a dream flowing over into the unconscious psychotic process, its content mixed with delusions and hallucinations.

How could the dream data be sorted into categories according to the format in which the violence was portrayed without reference to its specific content? Would it be possible, for example, to take dreams typed on cards and sort them into a few categories which represent how the work of the dream processes the traumatic event?

Classification of Dream Format

The *manifest content* was divided into three structural formats (Table II).

TABLE II.

Three Categories of Format of Catastrophic Dreams

Category		Absolute numbers	Percent of dreams
I	Actual	189	53
II	Variable	76	21
III	Hallucinatory	94	26
		359	100

Category I is the nightmare of the actual event as if it were recorded by camera verité. Many veterans spoke of these not as dreams but as facts and reality. No such dream was remembered as having occurred in combat. As a matter of fact, when asked if he had dreamed in Vietnam, the usual response of a veteran was incredulous, "Dream in Vietnam! in combat?" Even if they had dreamed such a nightmare, it is probable that on their waking to the reality-nightmare the dream paled and was forgotten. To not call them dreams because they are unlike any other dreams would be jerry-rigging data to fit theory. They are dramatic evidence of the reality of the psyche.

Category I dreams constituted fifty-three percent of the total dream sample. They are the only human dreams which define themselves in a completely predictable manner. They replay the same images of specific experiences again and again, recurring in a cyclic manner. They do not relate to the day's residue, the here and now, or pre-traumatic life experiences. They bear no relationship to transference which may partially explain psychoanalysts' general disinterest in them.

Category II dreams consist of war experiences that contain one or more sequences that conceivably might have actually happened but did not occur. In general these dreams are limited to the catastrophic events of violence but may contain images of the here and now or the dreamer's life experience which might have actually intruded into the event. These dreams lack the predictable nature of Category I dreams and the hallucinatory character of Category III dreams. Category II dreams accounted for twenty-one percent of the total. It is hypothesized that the dream images of the actual event are transformed by the psyche in processing the event into a psychological experience in metaphorical or symbolic ways.

Category III dreams are like ordinary nightmares, but their identification with the specific trauma or place of the trauma, i.e., the Vietnam war, is always present. Their metaphorical or symbolic quality is mostly self-evident. They frequently have images of the here and now and follow stimuli from the preceding day which can be clearly identified or inferred. But such dreams are obviously nightmares, and the story, or the set of images, could not have occurred in outer reality as is the case in all Category I and most Category II dreams. An illusionary and fantastic quality readily separates them from the other two categories.

It is hypothesized that in psychotherapy, or in spontaneous healing, Category I dreams are transformed into Category II dreams and then into Category III dreams, although in clinical practice one might encounter the change from I to III and not any intermediary stage. It is possible in observing these changes to understand the meaning of the experience in new ways. If this theory is correct, the emergence of an ordinary nightmare after prolonged recurrent reliving of the exact trauma in dreams is a healing process and, therefore, I call this process *the healing nightmare.* It is the psyche's attempt at healing. In therapy, transference enters into the Category III dreams.

Classification of Dream Motifs

Regardless of the format, it is possible to subdivide the *manifest content* of traumatic nightmares into broad but specific motifs. For the purpose of studying the Vietnam war dreams, the 359 dreams on cards were sorted into sequential narrative motifs by placing them in different piles designated as sequences in the narrative of war. These categories were not arrived at arbitrarily from imagination, but from a study of the dream content of individual dreams. The narrative or sequence was arbitrarily imposed to give it order. There is no such order in the dreams of individuals.

In therapy of other forms of violence, different narrative sequences might be inferred which are broadly anticipatory, experiential, and reflective. In the case of the Vietnam nightmares, there is an evolving story of war: anticipation, entry into zones of combat, approaching combat, engagement, alternate violent scenarios, return home, longing to return to Vietnam.

Dream motif is a term used to designate the dominant thematic quality of the dream. While there are images in dreams which relate to other motifs, it was possible to determine a *principal motif* in 319 dreams (Table III). To simplify classification, I arbitrarily limited multiple motifs to two. This occurred in forty instances.

The five ranking motifs were of war phenomenon in general, and the next four were principally dreams of killing. Attack, war, firefights were mainly Category I, while dreams dominated by the theme of the dead were equally divided between Categories I and III. When *death* is the overriding motif, the dead (viz., body count, corpses) are static in contrast with *killing and dying* motifs as images of action. Killing dreams are predominantly Category I. Metaphorical expressions of killing are by definition Category II or III. It is of interest that the dreaded experience of being captured and tortured, which was a reality for only one of these veterans, was predominantly Category II showing a strong reality hook, and little resemblance to ordinary nightmares. On reflection it might be said that the waking reality nightmare of *capture* was too real to be hallucinatory or fantastic, or rather that its reality was such. *Being*

TABLE III.

Dream Motifs Ranked According to Number of Dreams, Showing Also the Number of Dreams with Each Motif and the Three Format Categories

MOTIF CLASSIFICATION	DREAMS N	DREAMERS N	CATEGORIES OF DREAMS N I	II	III
1. Under attack	47	34	25	15	7
2. "War"	41	27	30	10	1
3. The dead	28	24	14	1	13
4. Firefights	27	22	17	10	0
5. Killing women and children	24	19	18	4	2
6. Metaphorical	23	16	0	4	19
7. Killing enemy	22	20	14	6	2
8. Killing buddies	22	18	20	1	1
9. Captured	19	11	1	14	4
10. Somewhere in Vietnam	17	13	11	4	2
11. Being wounded	16	12	7	5	4
12. Chase & running	16	12	4	11	1
13. Home	16	11	2	10	4
14. Being killed	13	11	0	0	13
15. Animals	13	10	3	7	3
16. Decapitation	13	9	0	2	11
17. Looming danger	12	10	10	0	2
18. Shot down	11	9	6	5	0
19. Atrocities	10	8	7	3	0
20. Return to Nam	9	8	0	4	5
TOTALS	399*	304**	189	116	94

* 40 dreams classified in two categories
** Approximately 3 dream motifs per subject

wounded is represented in all categories evenly. Dreams of *being killed* or *decapitated,* by definition, must be Category III dreams. Dreams in which others are decapitated are listed under *atrocity* dreams. In contrast to the emotion involved, atrocity dreams were limited to Categories I and II. There were few animal dreams.

No doubt the distribution of categories and ranking would vary with any sample. In the given sample there is a discernible pattern in the processing of the images from category to category.

Dreams of *killing others* were more frequent and more disturbing than dreams of being wounded or killed oneself. *Atrocity* and mutilation dreams ranked low, suggesting such events and images create less disturbance in the unconscious and more in haunting consciousness. The dark and archaic side of human nature presents affect-images of archetypes such as the shadow and even evil.[8] In war the archetypes of the hero and his battle with evil are dramatically experienced and exploited. Category I dreams may be the archetypal dream of war par excellence. Jung writes:

> The archetype is a kind of readiness to produce over and over again the same or similar mythological ideas. Hence it seems as though what is impressed upon the conscious were exclusively the subjective fantasy-ideas aroused by physical process. We may therefore assume that the archetypes are recurrent impressions made by subjective reactions. . . . Not only are the archetypes, apparently, impressions of ever-repeating typical experiences, but, at the same time, they behave empirically like agents that tend towards the repetition of these same experiences. For, when an archetype appears in a dream, in a fantasy, or in life, it always brings with it a certain influence or power by virtue of which it either exercises a numinous or a fascinating effect, or it impels to action.[9]

Examples of Dreams and Therapy

Examples of Category I dreams: "I see a chaplain giving last rites to a head." I did not make any interpretation, nor was it necessary. It is reasonable to assume that the rite was uniting body and spirit, a tacit demonstration that we are represented in our minds and our heads. If there is no body, the head serves as the body and gives a sense of ritual reassurance to the dreamer. Horror but reassurance. It is hard to imagine a chaplain giving last rites to a foot or a hand.

Some veterans' recurrent dreams are terrifying and phantasmagoric such as:

> I dream about my buddies. I can see blood pouring out of nowhere, and I see faces of friends of mine. Dead. I see battles going on and RPGs (rocket-propelled grenades) and tracers, and I hear explosions and see flames of napalm. Usually it is a lot of gory stuff. Dead. Mutilated bodies. Things like that. In the dreams I get terrified. They are in living color. I hear explosions and screams. Crying in pain. I don't feel it, but I sense the guys are feeling it. Once in a while I see guys I had contact with being hurt or killed. And dreams of getting even with the Viet Cong.

A specific event of violence appears in a recurrent unchanging dream:

> This happened at Meli Peak, Hill 101, where about fifty men were killed shortly after I arrived in Vietnam. I come in by helicopter. My buddy and I are about two feet apart in a foxhole on top of the hill. We get to know each other pretty good over three or four days. We have to stay in the same foxhole. We can't even get out to stretch our legs because of the sniper fire. I am talking with my buddy. It is dark. He lights a cigarette and all of a sudden his head blows off. His brains come out all over me. I wake up screaming.

Vietnam veterans were often labeled as baby killers. One veteran was haunted by this dream:

> I am on a patrol through a village, and this little kid five or six years old starts running toward us hollering, "GI! GI!," and holding up his hands. He has a grenade with the pin pulled in his hand. I shoot him before he gets us. I had to. He has more than one grenade strapped to his chest. He just blows up. There's nothing left of him.

Example of dream transformation in therapy: Jim is a thirty-six-year-old Mexican-American combat veteran who had a recurrent nightmare almost every night for twelve years before I saw him. At first I saw him every week for two or three one-hour sessions, and then once a week for six months, and less frequently for two years. I reported his analytic therapy in detail.[10] His nightmare was always about the same ambush:

> We are getting our gear together, and our briefing about the VC. I am point man for the platoon, and we go into a ravine. Suddenly there is shooting and yelling, and everyone is falling down and shouting or screaming in pain. There are seventeen men, another sergeant, and myself. The other sergeant and I get out of the ravine and blow away a VC in a foxhole and drag the guy down a jungle trail. I say, "I've gotta go back to the ravine." He says, "I ain't going back." I say, "I am." I jump into the foxhole and look down into the ravine. I don't do nothing because a VC officer is holding their heads by the hair and executing them one by one, and then he has shot them all and it's quiet and there is no more moaning. I throw a grenade and run back.

Jim had led seventeen newly arrived soldiers to their death. Nothing seemed to stop the dream. To cope with his agitation, insomnia, and nightmares, he was in and out of VA hospitals, resorted to drugs, alcohol, heroin, and large doses of psychotropic drugs. After two months the dream suddenly changed and, to his surprise and mine, he dreamed that he was the VC officer executing his buddies. Then his dreams became more hallucinatory:

> The whole platoon is after me. I can see the faces of the kids that got killed. They get me and tie my hands behind me and make me kneel on the ground. They put my head forward, and the sergeant cuts my head off with a sword. It rolls down the hill toward the ocean. After he cut off my head they let me go and I keep running and running, but I can't reach my head. I wake up in a cold sweat.

After the actual ambush, Jim made his way back to the base camp, and the next day they went to the ravine to get the bodies. There they found all the men had been beheaded and their heads put on punji sticks and smeared with shit. Some were mutilated.

After a series of dreams where I [Dr. Wilmer] appeared, the original ambush dream began to recur less frequently, and finally he said, "I still have the nightmare but I don't pay no attention to that bullshit."

Example of transference dreams:

> Everybody is getting killed. Harry, they got you in the center of the ravine bleeding from gunshot wounds. You are leaning against a tree and crying for help. I freeze and wake up drenched in sweat. *I change pajamas and go back to sleep because I want to finish the dream.* Now I am standing on top of the ravine, and a nurse is with me. She and I slip into the ravine. I hit the Viet Cong over the head with the butt of my rifle, and she dresses your wounds. I carry you out on my back. Sometimes in the dream your hands are tied. When I pull you out I wake up in tears. When it's you I cry. Before when I woke up it was always in a cold sweat but never crying.

His dreams had changed abruptly from Category I to III.

Jim was the hero rescuing the wounded healer. I appear in other roles in the dreams: "You were the commander of the battalion. The bodies were lying in the water, and you were screaming at me, 'Get out! Get out! This isn't for you! They're dead. Forget them.' "

A far cry from the nightly ambush.

Examples of Category II dreams: A veteran dreams that he pushes a Viet Cong out of a helicopter, something he had never done but had seen done. The obvious shift in the dream from the real event to what might have happened shows his unmistakable sense of guilt. He recalled:

> Just out of Chu Lai in a helicopter, a Vietnamese interpreter was trying to get the NVA (North Vietnam Army) prisoners to talk. And he told them

in Vietnamese to talk or we would throw him out. This was done by us. Still all in all it was something that was inhuman. He wouldn't talk. I don't know how high we were, but they kicked him out of the helicopter. He hit the ground. We were high enough to where I didn't even look back to see, but when we went down I saw he was literally split into four pieces.

One Vietnam veteran who had been involved in mutilation of prisoners' bodies and decapitation dreamed:

> I see heads hanging on a clothesline. There are five to ten heads. I am taking them off and holding them by their ears, reversing the way in which they are hanging. They are always oriental heads hanging by their left ear. Then I'd hang them by their right ear and move to the next head. What dawns on me in the dream was that there were teeth missing and that I am looking to see if there are any gold teeth left.

Examples of Category III dreams: "There is the stink of dead bodies. I am burying the dead—just the heads with big holes in them."

A medic tells me this dream:

> It is very hot. I am lying on my stomach in this field of grass. I can smell the pollen. I hear a guy crying, "Doc! Doc! Doc!," and I crawl through the grass until I come to the edge of the road. I have to get up and get him. I'm afraid. I get enough guts to go across the road to get him. As I do I get shot. It seems about fifty times, but I don't feel any pain. I just feel the thudding of whatever is hitting me. There's blood all over, but I still don't feel any pain. When I get over to where the guy is, I see that it's a buddy of mine called Cobb who was one of the first wounded guys I worked on in Nam. Actually he's been dead for a long time. In the dream though he is just looking up at me, and his face is all bloody. He says, "Welcome home!" I wake up drenched in sweat.

The dreamer comments, "I can't sleep. I'm afraid I'm going to have this dream which I've had over and over again. It never really happened. It's not even a realistic dream. The only thing that ever changes is their faces. Everything else is always exactly the same. One of the worst fears I had in Nam was that I would be afraid to go and get those guys who were wounded. But that was something that never happened. I always did my job. I never punked out going to get anybody. They gave me a Bronze Star for doing that. So I did my duty but in the dream I'm afraid. When I got shot in the leg up here [points to his left groin], when I came to I knew where I had been hit, but I couldn't feel any pain, and I was afraid to put my hand down to see if my leg was still there. Then I began to feel some pain. The next thing I felt was my crotch to see if it was still there. It seemed like hours because I was afraid to put my hand down there.''

A Special Forces Vietnam veteran began to develop nightmares twelve years after the Vietnam War was over:

> I am in a goddamn firefight. People are getting hit and firing back. Explosions are going off. Commands are being given. People are crying. We are tangling with the NVA in some small valley. We could see them coming over, and we seemed to be moving toward them. My damn weapon jams. I pull the damn cartridge out, whip it back in and start firing again. I pull it out, put in a new clip, but the automatic gun won't fire. By this time they are all over us. I look to the right and see a monk in a brown habit on the side of the mountain looking down. His hood slips down, and there's a skull, and it seems like it is laughing. I am scared shitless.

Example of dream therapy with nightmare: Identification with the aggressor is a well-known defense mechanism. Identification with the enemy, not as aggressor, but in a human role is a common dream experience. In a sense it is befriending the shadow. This often involves a return to Vietnam. Dream example:

> I am living in Vietnam and putting myself in the place of the Vietnamese people. I am a leader in the village. I see the Americans coming to attack; we had nothing to defend ourselves with, and I felt that they aren't going to injure us, but I get my people and lead them to escape from the village. We had attacked villages like this one, and I know how it was.

I asked him, ''Did the dream help?''

''No,'' he replied, ''it would put me on one side I saw. I've been on his side now. I know how the guy felt when he saw us coming because I took his side. I've been on both sides. It didn't help when I put myself in the place of the man who was leading these people out. I wanted to get out. I knew that if we stayed there we'd go back to the refugee camp at Saigon or Da Nang, or something. Because I'm a marine I know that we don't do those kinds of things, you know. In other words, murder people and stuff like that. They were civilians. I gathered the people, children, women and old men.''

It didn't *help* him to realize the fear and terror he had inflicted, but rather made him feel more helpless. He did not understand how I meant help as insight. The dream disturbed him. And yet because he denied atrocities in the name of being a United States Marine, it was his unconscious that made him question his role. How then to help him see deeper into the dream?

"Do you suppose," I inquired, "that there was any danger of their being hurt by the marines?" He tightened visibly and replied adamantly, "No. I don't think so. That's the way I felt."

"You could see both sides," I continued, "and learn to anticipate danger and get out. You could identify with the enemy, not as enemy but as other people, and see yourself as an invader and leader of the people. Of course, there were times when they were shot or killed."

He bristled. I had touched on the shadow and taken a chance in contradicting his feelings. It didn't work. He retorted angrily, "Well, this is my dream! This is the way I felt. This is what I experienced. This is me."

I had intruded. I was an invader. I had challenged his truth. But since I was neither angry nor hostile, and I understood his feeling, I went on, "Yeah, but you saw the others too, didn't you?"

"What?"

"Well, in other companies. I'm not saying that there's anything wrong with your dreaming. I think it's very good and very right."

He seemed relieved and sat back in the chair. Yes, maybe others. And he responded, "Right. Right. That's true. Because I saw a lot of the . . . (long pause) . . . I feel like the way you're saying it, because it is true."

Now I touched on his own heroism and courage. Changing the subject to help gain insight, I asked him, "How was it when you stepped on the mine?"

"Like a marine! Once a marine, always a marine!"

Once a man, then always a warrior, persona and shadow and heroic ego. While his collective projection on the marines was positive, so was his collective projection on the Vietnamese. By befriending the enemy, taboo to a personal marine, he was assimilating his shadow, and it was helping him to integrate his war experience. In this understanding he needed another human being. Neither the war experience nor his military identification was a block to me since I had been a Navy Captain during the Korean War, a positive experience, but we did not talk of these things.

Wish to Return to Vietnam

Sometimes a patient will awaken from a nightmare and want to go back to sleep to finish it, as in the case of Jim when he had left me bleeding and wounded, leaning against a tree in a Vietnam ravine. He wanted to go back to finish the action. In the case of Category III dreams, the return to the dream world allows the spinning out of the dream metaphors. The return to the Category I nightmare, in contrast, allows only more of a predictable horror. Sometimes veterans want to do this because of pleasure in the killing and a sense of power, but also it was consciously a wish to go back and finish the job right, to win the war, to beat the enemy, which of course never happened. Return to the war experience by vivid dreaming reinforced in some men a fierce sense of pride and masculinity, confrontation with death, and an illusion of invulnerability. One veteran expressed this desire in these words, "I've had a real urge, a bad urge again of wanting to go back to something that is over and done with and gone. Quite a few guys in units like mine still want to go back, especially ones that were wounded, because they figure they got a score to settle or they never got an opportunity to finish their job over there."

A veteran who spent two years as a door gunner on a helicopter reported this conversation, "My brother said, 'What's wrong with you? You just got back from Vietnam and you want to go back there.' 'I don't know. I just feel bored here. Over there you kept on going.' "

A segment from an analysis: A veteran whom I will call Sam sought my help but didn't want to be a part of the VA, so he saw me in my university office. He had a service-connected disability following extensive wounding when he stepped on a land mine. He was shy and spoke quietly and hesitantly. He tried to avoid being identified as a Vietnam veteran, and he had difficulty in trusting anyone, but over a period of eight months of weekly or twice weekly sessions he was able to trust me. His placid, controlled persona covered a volcano of affect and smoldering rage. He became particularly bitter when, during his work with me, a friend suggested that he apply for compensation as posttraumatic stress disorder which has just been recognized by the VA. To his humiliation, he was hospitalized for evaluation at another VA hospital and was diagnosed as malingering his nightmares, and a hostile report was put into his VA record. This was the only time during my study that I allowed my work to enter into the VA records, not in behalf of his compensation, but to contradict the accusation of malingering for this brave and depressed man. His usual dream was about the time of his wounding, but the night before his first session with me he dreamed,

> I am carrying a baby in my arms. I think it is my little boy. I don't

> remember seeing his face. I am inside this building trying to get shelter. We are under attack. I can hear the bombing, *Poom! Poom! Poom!* I am scared and running and running, carrying the baby. I don't recognize the other people. I don't know who is trying to drop bombs on us and kill us. All of a sudden I am running scared when I get hit. One of those things lands right next to me. I can feel the pain that the baby and I are going through. It is really real. I have this weird feeling like dying, man and the baby, and I say, "Oh my god, this is it." It seems like the real thing until I find myself shaking in bed. That is when I realize it is a dream.

The dream reminded Sam of the time when he was searching a hooch (native hut), and the instant he opened the door he heard a strange noise and jumped back thinking it was a booby trap. He said, "I didn't see it until it happened. A baby was lying on the cot." He began to cry. "I saw the baby. It still hurts. It was so burned. I knew the baby was inside the hooch. It wasn't my fault, was it?" Sam worried constantly that something might happen to his ten-month-old baby who was just beginning to crawl. He thought, "What is going to happen to him if I die? I'm still living the same nightmare [about the mine]. I told the physician when I was evacuated in Japan and he said, 'Don't worry. It'll go away.' OK. Then it never went away and they told me to see a psychiatrist. It doesn't happen every night. Every now and then it happens. When I go through this [Sam hits the table with his clenched fist] like when I got hit, I hate the booby trap. It scares me when it happens. When I don't have these dreams I'm OK. You know what I mean? They seem like the real thing. And I think when I dream like that, when they seem like they are the real thing, sometimes in the dream I say to myself, 'Is this a dream I am dreaming?' "

"In the dream itself?" I ask.

"Yeah. Getting worse. I just. . . ."

"Well, that's not a bad sign at all. That means that even in the dream you can question whether it is a dream, because in the most total dreams that question never comes up *in* the dream."

"I see."

"So that's a good sign."

"I used to think like that."

"It suggests that maybe you are half-asleep and you are kind of half-conscious, half-dreaming. Is that right?"

"Sometimes like that, right."

Sam's therapy took a positive turn around the time of a series of particularly violent episodes: the shooting of President Reagan; a berserk Vietnam veteran who opened fire from an RV (recreational vehicle) at a parade in San Antonio, carried live on television; and a series of gruesome murders in San Antonio. Dealing with this overt violence seemed to diminish the nightmare of the mine which subsided into relative obscurity. He remained moderately depressed but was able to handle his war trauma with more control, strength and understanding.

The question "Am I dreaming?" means becoming more conscious. And his initial dream anticipating his first visit to me was already a transformation of the Category I dream (the mine) to a Category II dream (the baby), symbolically the one (also) booby-trapped, as well as his actual child and himself. That major shifts can occur just prior to a significant consultation has been observed many times: I have reported an instance of a patient whose disabling symptom disappeared right after mailing his letter to me for an appointment.[11] The spontaneous healing of symptoms and such symbolic transformations in dreams suggest the psyche's processing mechanism—*vis medicatrix naturae,* the healing power of nature.

Commentary

> "Merlyn," said the king, "It makes no difference whether you are a dream or not, so long as you are here. Sit down and be patient for a little while, if you can. Tell me the reason for your visit. Talk. Say you have come to save us from this war."[12]

There is something intrinsically shattering in hearing and being with a survivor of a catastrophe as he tells one his story. The shattering factors appear to be that the listener is unconsciously involved in the horror which evokes the therapist's own shadow, and awakens feelings of his own violence, murderous impulses, primitive thoughts and feelings.[13] Of course, it is possible for the listener to deny or avoid listening to the pathos, and thereby be neither disturbed nor involved—not even there really. Although evil is a part of war, there are also great acts of heroism, altruism, bravery and sacrifice. War also brings out the archaic savage that still lives within us all. Listening arouses in the therapist a moral agony of good and evil, black and white, healer and destroyer, right and wrong, with conflicted feelings such as attraction and revulsion, and hence intense countertransference. The therapy of violence needs bonds such as dreams provide.

C. G. Jung wrote:

> We could say without too much exaggeration, that a good half of every treatment that probes at all deeply consists in the doctor's examining himself, for only what he can put right in himself can he hope to put right in the patient. It is no loss either if he feels the patient is hitting him, or scoring off him: it is his own hurt that gives the measure of his power to heal. This, and nothing else is the meaning of the Greek myth of the wounded physician.[13]

Hence the admonition, Physician Heal Thyself.

In psychotherapeutic interviews with survivors it seems wise to be reminded of Novalis's aphorism, "Every word is an exorcism. Whatever the spirit that calls, a kindred spirit will answer."

No matter how we look at catastrophic violence, we are of two minds about it. When the listener hears the story of the trauma, it is literally both in the mind of the teller and the listener; and while in the latter it resembles the former, in each it is perceived according to his or her training, experience and nature, rather like the breaking of the granite.

The collective feelings and symbolic legacy of the war are entombed in the Vietnam war memorial in Washington, D.C. There, over 58,000 names of Vietnam veterans killed or missing are engraved in 400 feet of polished black granite. When you look closely, you see your own dark face reflected back at you coming from behind the names of the dead.

> *Facilis descensus Averno:*
> *noctes atque dies patet atri ianua Ditis;*
> *sed revocare gradum superasque evadere ad auras,*
> *hoc opus, hic labor est.*
>
> The gates of hell are open night and day;
> Smooth the descent, and easy the way:
> But to return, and view the cheerful skies,
> In this the task and mighty labor lies.
>
> (Vergil, *Aeneid* 6. 126)

1. M. W. Brown, F. E. Williams, eds., *Neuropsychiatry and the War: A Biography with Abstracts* (New York: War Work Committee for the Mental Hygiene Association, 1918). E. E. Southard, *Shell-shock and Other Neuropsychiatric Problems, Presented in Five-Hundred and Eighty-One Case Histories from the War Literature, 1914–1919* (Boston: W. M. Leonard, 1919). A. Kardiner, *The Traumatic Neuroses of War: Psychosomatic Medicine Monograph* 2–3 (Washington, D.C.: National Research Council, 1941), pp. 88–95. A. Kardiner, "The Traumatic Neuroses of War," in *The American Handbook of Psychiatry,* ed. S. Arieti (New York: Basic Books, 1959), pp. 245–57.

2. M. Kramer, L. S. Shoen, L. Kinney, "The Dream Experience in Dream-Disturbed Vietnam

Veterans," in *Post-Traumatic Stress Disorder: Psychological and Biological Sequelae,* ed. B. A. van der Kolb (Washington, D.C.: American Psychiatric Association Press, 1984), pp. 82–95. H. J. Schwartz, ed., *Psychotherapy of the Combat Veteran* (New York: Spectrum Publications Medical and Scientific Books, Inc., 1984), xi–xxvii. H. Hendin and A. P. Haas, *Wounds of War: The Psychological Aftermath of Combat in Vietnam* (New York: Basic Books, 1984). R. Blitz and R. Greenberg, "Nightmares of the Traumatic Neuroses: Implications for Theory and Treatment," in *Psychotherapy of the Combat Veteran,* pp. 103–23. V. J. DeFazio, "Dynamic Perspectives on the Nature and Effects of Combat Stress," in C. R. Figley, ed., *Stress Disorder among Vietnam Veterans: Theory, Research, and Treatment* (New York: Brunner/Mazel, 1978), pp. 36–38.

3. E. A. Brett and R. Ostroff, "Imagery and Posttraumatic Stress Disorder: An Overview," *American Journal of Psychiatry* 142/4 (1985) : 417–24.

4. H. A. Wilmer, "Post-Traumatic Stress Disorder," *Psychiatric Annals* 12/11 (1982): 995–1003.

5. H. A. Wilmer, "What Do You Say When a Patient Tells You a Dream?" *Texas Medicine* 78/4 (1982) : 46–48. J. A. Veninga, "The Healing Nightmare: A Conversation with Harry Wilmer," *Texas Humanist* 5/2 (1982) : 3–11. H. A. Wilmer, "Dream Seminar with Schizophrenic Patients," *Psychiatry* 45/4 (1982) : 351–60. H. A. Wilmer, "War Dreams: A Decade after Vietnam," in J. A. Veninga and H. A. Wilmer, eds., *Vietnam in Remission* (College Station, Tx.: Texas A&M University Press, 1985), pp. 90–106. H. A. Wilmer, "The Healing Nightmare," *Quadrant* (spring 1986), in press.

6. *CW* 8, §474.

7. T. C. Chamberlin, "The Method of Multiple Working Hypotheses" (reprinted from *Science,* 1890), *Science* 148/3671 (1965) : 754–60.

8. A. Stevens, "Shadow: The Archetypal Enemy," in *Archetypes: A Natural History of the Self* (New York: William Morris, 1982), pp. 210–43.

9. *CW* 7, §109.

10. H. A. Wilmer, "Vietnam and Madness: Dreams of a Schizophrenic Veteran," *Journal American Academy of Psychoanalysis* 10/1 (1982) : 47–65.

11. H. A. Wilmer, "Transference to a Medical Center: A Cultural Dimension in Healing," *California Medicine* 96/3 (1962) : 173–80.

12. T. H. White, *The Book of Merlin* (Austin, Tx.: University of Texas Press, 1977), p. 11.

13. *CW* 16, §239.

Acknowledgments

Supported by grants from the Rockwell Fund, Houston, and the Harris Foundation, Chicago.

All of the audiotapes and all of the transcriptions of the dreams, interviews and therapy have been placed in the archives of the Humanities Research Center at the University of Texas at Austin.

All names of veterans are deleted. I am grateful to Dr. Thomas Newcomb, formerly Chief of Staff of the Audie Murphy Veterans Hospital at San Antonio, for making this study possible.

PSYCHIC WOUNDS AND BODY SCARS
An Exploration into the Psychology of Keloid Formation

ROSE-EMILY ROTHENBERG
(Pacific Palisades)

Clinical Aspects of Keloid Formation

Scars are imprints of traumas once present and now left behind, vestiges that serve as distinct reminders of the injuries that body and psyche have encountered. Ordinary scar tissue, at first red and raised above the surrounding skin, fades and begins to shrink slightly with age. Keloids, by contrast, result from a proliferation of scar tissue and grow beyond the site of the original injury. Keloids are autonomous overgrowths of scar tissue. Elevated, irregularly-shaped, progressively enlarging scars, they develop a shiny, smooth, taut surface and a hard consistency due to excess collagen formation. Keloids evolve as a result of a trauma; the initial wound can be as extensive as a severe burn or cut, or as trivial as an acne lesion.

The first clinical description of the keloid was made by Louis Alibert in 1814.[1] He originally called it "cancroid," later changing the name to "cheloid" or keloid. Its name is derived from the Greek word meaning "claw"; extensions into the neighboring tissues often resemble the claws of a crab.

Keloids are generally one specific growth but may occur in multiple tumors varying in size and shape, mostly located on the upper back, chest, breasts, earlobes and on abdominal incisions. They are benign but produce irritating symptoms such as burning, itching, pulling sensations and inter-

---

Rose-Emily Rothenberg is a Jungian analyst practicing in Pacific Palisades, California. Her special interest—on which she lectures and gives seminars—is the relationship between disease and the psyche. This paper was first presented at the 1985 Joint Conference of Jungian Analysts at Carmel, California.

mittent smarting pain, and are generally disfiguring. Keloids are more common in dark-skinned races with a higher incidence in younger individuals, occurring more frequently during times of puberty and pregnancy, perhaps from hormonal influence. Treatments to remove or reduce keloids have met with variable success. The most common are surgery, radiation, steroid injections and pressure devices, though surgery is often avoided because relatively small lesions may become extensive, deforming ones as a result of attempts at removal. They are persistent; "like the Hydra of Greek legend, the keloid grows two heads each time one is excised."[2]

Researchers have defined structure, function and biochemical factors in keloid formation, but their etiology is unknown. Though true keloids rarely arise de novo, it is my hypothesis that psychic injuries can result in keloid formation. My own experience with keloids indicates that they do emerge on the skin in a "spontaneous" fashion and that psychological wounds may be the precipitating factor. Psychic experiences can generate physiological effects including hormonal imbalance and immunological deficiencies; persons predisposed to keloid formation can produce a lesion resulting in keloid growth as a consequence of psychic experiences that are traumatic and wounding.

Keloid Formation in Prehistoric Artifacts and Tribal Customs

When I found that keloids are artificially produced in primitive tribes, it confirmed my suspicion that there is indeed a primordial image of the keloid in the collective unconscious making its way into the world via the human body. Till my late twenties, my only experience of keloids was my own scars and a few unsightly pictures of very large keloids in some medical books. Then I came upon a book, *The Art of Mesopotamia,* with photographs of figurines found in a fourth-millennium grave in southern Mesopotamia.[3] There were little pellets of clay on the figurines' chests, shoulders and upper arms located where my own keloids are. Scholars have been unable to determine the exact significance of these artifacts, but because they were associated with the dead and presented features emblematic of the Gods of the underworld, they were viewed as chthonic deities[4] and because the figurines are found on the pelvis of the dead, they were thought to symbolize fertility and life.[5] The clusters of "knobs" on their bodies are thought to be cicatrices.[6] Figurines with these scar-like markings dating from the end of the seventh millennium B.C. were found in Achilleion, Thessaly, northern Greece.[7] Similar pellets of clay appeared on sculptures found in far west Mexico in tombs of a civilization that flourished around 1800 B.C.[8] Archeologists have also found Nigerian terracotta heads from the Nok civilization dating from the ninth century A.D. that depict keloidal lesions.[9]

The art of body marking is an age-old custom most commonly found in African societies and among the aborigines of Australia and Polynesia. Legends regarding the origins of body markings are preserved in oral accounts. One Australian tradition relates that scarification was taught the natives by a spirit who was then transformed into a kangaroo.[10] One Andamanese tale, describing scars as sacred marks of protection, tells how the mythic ancestor of the Mincopies was the first to have marked himself with scars. After tattooing himself he sang, "Who can kill me now? I am tattooed, I am tattooed."[11] A Nigerian legend tells of a man with two wives, one of whom he loved more than another. One day a crocodile was caught and placed in the room of the wife he didn't love. She admired the raised geometrical patterns on his back so much that she took the crocodile to a blacksmith and asked him to cut the same patterns on her body. After the operation, the markings were so engaging that other women began to copy them.[12] That it was the unloved wife who receives the markings suggests to me that dealing with her dark state of affairs results in transformation; her difficult circumstances mark her but also give her an individual status.

Scarification is performed on specific occasions. In initiation rites (the most commonly known rituals associated with scarification), it is frequently accompanied by circumcision, subincision, tooth extraction, etc. These ceremonies celebrate transformation, rebirth, and the continuity of life. As part of the ceremony, the initiate commits him/herself to upholding the social patterns and beliefs of the group. In this new relationship to the collective, the individual carries more weight and is a more active participant in the group. The mystery involved in the ceremony, preparation, period of solitude, and splendid festivities following the scarifications testifies to the event's sacredness. For this to have a full and lasting effect on the initiate, the body is ritually involved. For the primitive psyche, body decoration is vital to any celebration.

Those who practice scarification believe that lessons will not be forgotten if inculcated at a time of extreme emotional tension and stress; for this reason body operations performed at initiations are paid for in pain and blood. They help the initiate develop courage and endure suffering without complaint, and scars become permanent proof of his/her fitness and endurance. The shedding of blood is thought to be more tribally binding than any oath. Scarification is an indication of the initiate's new status among his fellow tribesmen.

In the Nuba, every female undergoes several scarification rituals, the first when she is ten or eleven years old, the second some time after the beginning of menstruation.[13] Subsequent scarifications are made when she expects her first child and after it is weaned. She may receive a new design with each pregnancy. These scarifications have an erotic component and are associated with fertility. The more three-dimensional the scars become, the more desirable she becomes. The process of actually making the keloid scars begins with rubbing oil onto the parts of the body to be incised and tracing a line where the incisions will be made. Then a portion of the flesh is opened with a thorn or knife and an irritating juice is injected—or ash, earth or charcoal is rubbed into the wound, allowing it to swell. Some wounds are opened many times to enlarge them.

When the incisions have been made and the irritant injected, the wounds are dusted with a powder of roots and herbs which helps to ease the pain and protect the wound from infection. The rite has its risks, even beyond infection. Disfigurement can develop, and a woman can lose a great deal of blood during the operation. Some lose consciousness. But to the initiate, the end result is worth the price as the scars render her highly attractive and, in effect, a sacred object.

An example of male initiation rites concerns the aborigines of Australia.[14] I will describe only those parts of the ceremony having to do with scarification. In these secret rites for men, the initiates go through three steps, each representing an epoch in their societal life. At the age of fourteen or fifteen, they enter the first stage. Little is known of this ceremony except that the novices are required to paint their faces black and speak in whispers. They advance to the second degree and undergo circumcision. The last and most important ceremony is at eighteen or twenty years of age when scarification takes place. It is at this point they are given names that they will henceforth bear. Each name must be totally new, never having been used by any other native, alive or dead. Their bodies are scarred with quartzite fragments. At the climax of the ritual, several of the older men open veins in their own arms, and the initiates swallow the first drops of blood. Next they prostrate themselves, and their backs are then covered with a thick coat of blood which is allowed to congeal. One man marks with his thumb the places on their backs where the incisions are to be made. Though great pain is inflicted, the initiates do not utter a sound or move a muscle. Drumming and singing often accompany these ceremonies, as a distraction for the initiate and to encourage him to endure the pain.

The placement of the scars will carry meaning, varying from group to group. In the Nuba, when the girl undergoes cicatrization at the onset of her first menstruation, the design consists of little dots cut on the upper part of her chest and breasts.[15] In the Tiv, a society in central Nigeria, the abdomen may be scarified when a girl reaches puberty or when she expects her first child. It has been suggested that the lines in the abdominal designs may represent the woman's genealogical lineage, her ancestral land and tradition passed on from generation to generation. The circle may represent the ripples on the pond with the past, present, and future circling around the navel as axis. The scars represent fertility in terms of human life and in terms of agriculture; the adult Tiv woman bears the responsibility for the crops. The designs may also serve to mark the woman's generation since the patterns themselves are changed every ten years or so according to current fashion.[16]

Some writers have reduced scarification to attempts to prevent or heal disease. As a form of inoculation, the cuts were believed to produce immunity and to aid the healing of injuries. Inflaming the wounds with dust and bacteria would heighten the body's powers of resistance since the body develops antibodies that make later injuries heal more quickly. Sickness was often thought to be caused by spirit possession, so the bloodletting aspect of scarification made it possible for the imperfections or the evil contained in the body to escape. Scarification in this view was primarily preventive medicine, protective in purpose.[17] Another theory about the function of body markings derives from slave trading. Facial marks, widespread among many Nigerians, have been used to identify family members sold into slavery because the scars also serve as family marks.[18] In West Africa, every group or clan within a group had its own skin pattern. For the Yoruba and Nupe, the use of marks enhanced the children's resemblance to their parents and thus clarified legitimate ownership of the children.[19]

In addition, keloids are thought by some authors to have a highly religious connotation. In Polynesia the deity to whom the person dedicated himself could be identified by the patterns of the scars. These permitted the God to descend on a man and enter him, leaving the God's sign upon the man's body. The God tolerated no one in his environment who did not carry his sign.[20]

Production of scars often accompanied mourning. Tattooing was a way to bring to memory someone deceased. Self-wounding or flagellation was also used when a relative died; aborigine women show their grief by gashing themselves quite severely in mourning for the dead, scoring themselves with the sharp end of a digging stick or club, or beating their heads with heavy stones. After gashing the body, one customarily rubbed sand or ashes into the wound so that huge keloid scars resulted.[21] In some African societies, slaves and their wives were scarified on the death of an important chief. They cut themselves until blood flowed to convey proper grief to the spirit hovering nearby.[22]

Creating these scars as art forms keeps alive the myths and philosophies of the tribe, as well as the connections to the tribal ancestors. Every design is different in some way in order to permit the spirits to reach each person. It is one's uniqueness, as represented by individual scarification, that ensures recognition by the Gods; the scars serve as antennae to receive their cosmic waves. It is also said that the keloids are created to make additional space for the spirits to reside. The Nuba believe that without their distinctive scars they will not be admitted by their ancestors in the aftertime. The keloids are the messages of their forebears; the marks on the skin are their words.[23]

The broad spectrum of meaning behind scarification ceremonies illuminates the keloid archetype. For primitive peoples, keloids carried the numinosity intrinsic to events of initiation. They also carried the significance of personal and collective identification, as well as the need for protection and a living relationship to the spirits in the hereafter. The keloids have carried a corresponding numinosity for me, a twentieth-century person of Western civilization. Beyond that, the appearance of my keloids has striking parallels to the events of the scarification ceremonies in the preceding material.

A Personal History of Keloid Formation

Spontaneous keloids and those induced through medical intervention arose in my life during initiations that were accompanied by emotional stress and pain. My first keloid grew out of a vaccination I had when I was about six. (Vaccinations were required for admission to public school.) A mole was removed from my other arm at the same time; it also became a keloid. Both were surgically excised when I was eleven but grew back larger than before. When I went away to college, a rash of keloids appeared across my chest. When I married, new scars formed an epaulet on my right shoulder. When I gave birth to our child, a keloid appeared on my breast.

My clinical history reveals that their emergence was associated with events of initiation that were particularly stressful. A rite of passage involving the birth of the new and death of the old inevitably creates a wound. If the wound is very deep, considerable healing is necessary. This was so for me. My initial wound was my mother's death when I was six days old. When I was two—long before I became conscious of that painful fact—a stepmother came on the scene. At six when my scars emerged, I thought of them as the third injustice I had to bear. I believed that for many years, but through this work I see now they served to protect me from all I had to withstand and that I owe them a great debt.

A number of circumstances bear directly on my receptivity to keloid formation. In my childhood several books touched me very deeply. They may have prefigured my skin condition or may have influenced its development. My favorite book from earliest years was *Wee Fishy Wun,* a story about a little black fish that turned into a beautiful goldfish.[24] Tiny black beads were attached to the fish in the first part of the book; they created a nubby effect you could feel. On the last page, tiny, glistening gold beads decorated the fish in place of the black ones. I especially loved touching the little beads on the goldfish. Perhaps I was identifying with the longed-for transformation process they represented.

A second book was a favorite of my father's (it came from a collection of books he shared with my real mother) and became a favorite of mine. He used to read and reread this book to me as I sat on his lap. It was called *Gods' Man* and was composed of a series of woodcuts.[25] This was not a book intended to be read to a child, but his sharing something from my mother's era made it special for me at the time. The story was very dark, but I did not connect with this until my later years. The woodcuts told of an artist's search for his identity. In a state of poverty and innocence, he signs a pact with the devil who offers him money and professional success. Later on an entrepreneur buys his paintings and sets him up in a studio with a mistress. The turning point comes when he discovers she has a dollar sign tattooed on her shoulder and he realizes the true nature of her seductiveness. He is devastated, and a litany of humiliating experiences follows. After a brief respite with a wife and child, the devil comes to claim him.

My father must have identified with the destiny of the protagonist. When my mother died, his spirit seemed to die too. Feeling guilt for taking her away, I was susceptible to carrying my father's projected anima aspects—the wife, the whore and his unlived creative life. The arrival of the keloids seemed to offer me a protection and distraction from this heavy burden. Much energy went into getting treatments for my scars and protecting me from injuries that might lead to further scarring. I was also experiencing an entangling relationship with my stepmother, becoming the target of her frustrations and disappointments in being second wife and second mother. Again the keloids offered a protection from her attachment to me. When the scars appeared, my stepmother was visibly uncomfortable looking at them; if I undressed in her presence she hurriedly left the room. The keloids served as a defense for me, as the image of the crucifix did for the medieval hermits who held it up to ward off the demons of darkness.

I developed a sizable complex about hiding my scars. When we went shopping, my stepmother kept the salesladies out of the dressing room to avoid their asking questions about them. Changing clothes in the girls' locker room was a nightmare. The guilt and hurt that belonged to my orphan psychology were projected onto the scars. Years later when I was in analysis, I wrote what it was like: "I live in dread that my scars might be exposed. I am ashamed of them. I am disgraced that I am different. I am like a leper and people won't want to be with me and I'll be left all alone again. My mother's body left me. I am angry at my body. If it had been good enough she wouldn't have left. This imperfect body of mine has gone and betrayed me with disfigurement. These scars are the very pain of my upbringing: the shame I feel of my stepmother and my connection to her is the same as the shame I feel about the scars. The two are intertwined. The fear I have of her is the same fear I have of showing my scars."

In *Death in Life,* Robert Lifton wrote of some survivors of Hiroshima who developed keloids as a result of their burns and felt guilt, shame and self-condemnation in the face of the others' deaths.[26] Their scars, visible reminders of the horror they had experienced, were considered marks of disgrace. Despite the enormous difference between their experience and mine, the feelings weren't wholly alien. Lifton interviewed one of the victims:

> I have a special feeling that I am different from ordinary people . . . that I have the marks of wounds—as if I were a cripple . . . that I am inferior to them . . . of course physically, but also mentally. . . . Ordinary people don't have this kind of scar. They don't have to experience the feeling of humiliation that I have had. . . . I imagine a person who has an arm or a leg missing might feel that same way. . . . It is not a matter of lacking something . . . but rather . . . a handicap—something mental which does not show . . . the feeling that I am mentally different . . . and incompatible with ordinary people.[27]

Lifton continues:

> The stigmata can assume affirmative significance—on the order of what Saint Paul called "marks of the Lord Jesus on my body." But more often these ennobling associations are weak or absent. The keloid then comes closer to the pre-Christian meaning of the stigma as a mark or brand for slaves and criminals, and to the related "post-Christian" idea . . . "A mark of disgrace or infamy." Most of all, it takes on the recent medically influenced meaning . . . an "indication of disease" in which the "disease" mars not only the bodily surface but the entire idea of the Self.[28]

I too have experienced my keloids as acts of God. Because of difficulties in early life, I equated them with the dark side of the Self. I felt like an imprisoned victim, and they became my 'brands of slavery.' At the same time they embodied the protective spirits; when I was able to make a conscious relationship to this aspect, they provided me with a tangible experience of the positive side of the Self.

The transformative aspect of the keloids became visible to me in a dream I had in college: "I had on a black evening gown with one shoulder strap. The many scars I had on the opposite side of my chest and arms formed flowers, and the total effect created the other strap." When I awoke I felt a great sense of relief and gratitude that the scars could become a decoration instead of a disfigurement. I knew nothing of scarification rituals at that point. Around the time of the dream, I had participated in a faith-healing ceremony to arrest a keloid that was developing from a cut on my knee. After the ceremony, the raised scar formation dissipated and the keloid failed to materialize. I discovered then that I had some control over the keloids through a receptive attitude to a healing spirit. I felt like Job in needing to enact a more conscious relation to the darkness within myself and to the dark forces without. I once did a collage of my dream image and added a multitude of black flowers scattered about my body. Those flowers represented the dark aspects in my psychology I needed to work on, one by one. My scars were my Scarlet Letter, a visible admission of my sins. (In preparing this paper, I find that the black [rose] flower symbolizes the silence of an initiate.)[29]

Several years ago I dreamed: "A man was standing on the sidewalk corner talking to some women about his keloid that had grown where his penis had been removed. He had a new penis in his hand. It was large and powerful. I was passing by and overheard the word keloid as he was proudly showing it, and wanted to tell him that the penis has the thinnest skin of any area of the body and that's why it is the most sensitive. He thought that was very interesting." In this dream the keloid is the replacement for the lost penis. It is a substitute for the creative organ, the principal fertility symbol. Osiris's penis was swallowed by a fish, and a wooden one was made by which he conceived Horus. The keloid's connection to the penis represents its function of transforming libido, the death and rebirth of creative energy. The sensitivity may refer to my own sensitivity about my scars.

In another dream, I was reading proofs for my book in the medical building where I went with my stepmother for treatments for my keloids. The building had become Bollingen Press! This remarkable transformation indicated that healing was taking place, and it related to my writing on the subject of keloids. A relevant association came to me when I wrote about this dream. I remembered that, when I was in the doctor's office, he showed me some awesome pictures of large keloids on a woman's ear resulting from an ear piercing. At the same visit he spoke in whispers to my stepmother about the possible dangers of my becoming pregnant: scarring could result from caesarean section or episiotomy. With this memory another dimension to my scars unfolded—my father told me once he was afraid no one would marry me because of my scars, and I remembered that the high school counselor discouraged me from going to college since my SAT scores were low. From these memories I speculated that the emergence of my scars at college, marriage and childbirth could have been my body's reaction to these negative predictions, saying: "It's not true and we'll show them. We'll rise above it!" My scars continue to speak to me in active imagination. They connect me to my primitive roots and have told me they were related to and were replacements for my real mother, that I could feel secure; they were permanent fixtures and would not leave me.

The last dream I wish to include dramatically portrays the keloid archetype: "A close friend of mine, whose relationship to me was very healing, was eating my scars and they were grapes." The highly creative animus this male friend represented was eating my scars and digesting what they represented. The grapes are a connection to Dionysus, and keloid scar tissue formation is in its plenitude or excess like a Dionysian orgy. In *The Mythic Image* Joseph Campbell has included a picture of "Dionysus as the Deified Grape," indicating its transformation aspect: "the imminent divinity that had been there in the grape all along."[30]

Symbolic Aspects of Keloid Formation

Purposely made, keloids are recognized as body art; seen in the doctor's office they are classified as skin disease. Unbidden or not, they belong to the Gods. From their beginnings in antiquity, they seem to have been a numinous entity defying the physician's scrutiny. In 1790 Retz wrote what is probably the earliest description of a keloid in a "Treatise on Skin Diseases and Things of the Mind":

> The epidermis is not altered: but is often a dark red colour. It is raised up by a mass of solid material which may be as large as a nut or apricot, or as long finger-like processes, thick like that Italian dish called macaroni. These plaques . . . may look like scars with folds and puckers, as though there had been many scars one upon another.

(The comparison in clinical description with food and drink was and is still common in medical literature.)[31]

In modern dermatology texts, keloids are not mentioned among skin diseases with a probable psychogenic component. Yet a few medical researchers and writers have been conscious of the possibilities of such a relationship. One physician expresses his wonder at the keloid's recurrence: "It seems so reasonable to excise [keloids] and so unreasonable that they should return bigger than ever."[32] A subtitle in a French article on keloids suggests the writers' awareness that the origin of the keloid may be connected to some aspects beyond the merely physical: "The Undesirable Skin Scar: Punishment or Destiny?"[33] Another dermatologist makes this concession to the subjective element: "Although keloids do have a life cycle, it is one measured in decades. Hence, no one can expect our patient to view her progressively disfiguring scar as a self-limiting spontaneously resolving process. She needs treatment but assuredly not surgical!"[34] And a Nigerian physician writing on the origins and history of Ifa divinations, a cult of the Yorubas, appreciates the role of divine intervention: "They [Ifa priests] also knew that once a lesion appears it grows in size and has no remedy except when 'the Divine power is suitably appropriated to intervene in bringing about its resolution.' "[35]

To bring about a significant change in one's life, some degree of suffering seems necessary. For keloid-formers the suffering is often both physical and psychological. Intermittent pain and itching are constant reminders of the keloids' presence and what they may represent. That the scars come unbidden, are unattractive and cannot easily be removed is an added element of distress. Whether a mark of beauty or an ugly scar, the keloid is a reminder of wounds and pain once experienced and the associated sacrifices these wounds may have required. It is produced from a state of tension and, chosen or not, carries a high price.

Within the keloid the divinity resides, a living symbol of the transformation process. Jung said, in writing about layers of the skin as the protection against outside influences: "The same motif is expressed by the petals of the lotus and by the skins of the onion: the outer layers are withered and desiccated, but they protect the softer, inner layers. . . . In Christian metaphor, Mary is the flower in which God lies hidden. . . ."[36]

The use of the keloid to represent and bring about transformation underlies this powerful archetype. The skin protuberances have been compared to warts on the toad and frog, animals that symbolize transformation. The frog and toad often represent the despised and rejected parts of ourselves; accepting them has a transformative effect. As Shakespeare says in *As You Like It:*

> Sweet are the uses of adversity;
> Which, like the toad, ugly and venomous,
> Wears yet a precious jewel in his head.[37]

Persons scarring themselves for ritual reasons know the meaning behind the keloid and seek to make it concrete. For the modern patient who suddenly finds he/she has a scar that defies medical intervention, the meaning is not so readily apparent. Can the same process belong to both? In each the archetype seems to push upward into its keloid form and bears the numinous power of its mark. Because the keloid carries such power and numinosity, it is desired by the primitive soul. The keloid is an aberration, and the aberrant is sacred. The keloid's connection to the divine may have motivated the primitive psyche to create it; for the modern-day patient, alignment with the divine may offer a way to bring about its resolution.

1. David Bloom, "Heredity of Keloids: Review of the Literature and Report of a Family with Multiple Keloids in Five Generations," *New York State Journal of Medicine* 56/4 (Feb. 1956): 511.

2. Alibert as quoted by Walter B. Shelley, *Consultations in Dermatology with Walter B. Shelley* (Philadelphia: W. B. Saunders Company, 1972), p. 282.

3. Eva Strommenger, *The Art of Mesopotamia* (London: Thames and Hudson, 1964), plates 10, 11, 12.

4. Sir Leonard Woolley, *Ur Excavations: The Early Periods,* vol. 4 (Philadelphia: The American Philosophical Society, 1955), p. 13.

5. Yasin Mahmoud, "Clay Figurines in the Iraq Museum" (Master's thesis, Dept. of History and Archaeology, University of Baghdad, 1966), p. 233.

6. Woolley, *Ur Excavations,* p. 12.

7. Personal communication to the author; photocopies of two figurines with pellets from Achilleion, Thessaly, northern Greece, dated from the end of the seventh millennium B. C. given by Marija Gimbutas.

8. Jacki Gallagher, *Companions of the Dead: Ceramic Tomb Sculpture from Ancient West Mexico* (Museum of Cultural History, UCLA, 1983), p. 18.

9. P. Omo-Dare, "Yoruban Contributions to the Literature on Keloids," *Journal of the National Medical Association* 65/5 (Nov. 1973): 371.

10. Theodor Waitz and Georg Gerland, *Anthropologie ber Naturbolfer* (Leipzig: Friedrich Fleischer, 1872), p. 740.

11. Ernst Grosse, *The Beginnings of Art* (New York: D. Appleton and Company, 1897), p. 73.

12. Corneluis Oyeleke Adepegba, "A Survey of Nigerian Body Markings and Their Relationship to Other Nigerian Arts" (Ph.D. diss., Dept. of Fine Arts, Indiana University, 1976), p. 18.

13. Leni Riefenstahl, *The People of Kau* (New York: Harper & Row, 1976), p. 221.

14. R. Brough Smyth, *The Aborigines of Victoria: With Notes Relating to the Habits of the Natives of Other Parts of Australia and Tasmania* (London: Trubner and Co., 1878), pp. 66–68.

15. Victoria Ebin, *The Body Decorated* (London: Thames and Hudson, 1979), p. 56 (fig. 58).

16. Bruce Lincoln, *Emerging from the Chrysalis: Studies in Rituals of Women's Initiation* (Cambridge: Harvard University Press, 1981), pp. 46–49.

17. Leni Riefenstahl, *The Last of the Nuba* (New York: Harper & Row, 1973), p. 18.

18. Adepegba, "Survey of Nigerian Body Markings," p. 29.

19. Ibid., p. 39.

20. Waitz and Gerland, *Anthropologie,* p. 37.

21. Stanley D. Porteus, *The Psychology of a Primitive People: A Study of the Australian Aborigine* (London: Edward Arnold and Co., 1931), p. 121.

22. Isaac F. Marcosson, *An African Adventure* (New York: John Lane Co., 1921), p. 195.

23. Andre Virel, *Decorated Man: The Human Body as Art* (New York: Harry N. Abrams, Inc., 1980), p. 132.

24. Kira Melissarato, *Wee Fishy Wun* (New York: Cupples and Leon Company, 1945).

25. Lynd Ward, *Gods' Man* (New York: Jonathan Cape and Harrison Smith, 1929).

26. Robert Jay Lifton, *Death in Life: Survivors of Hiroshima* (New York: Basic Books, Inc., 1967), pp. 178–79.

27. Ibid., pp. 177–78.

28. Ibid., p. 181.

29. Gertrude Jobes, *Dictionary of Mythology, Folklore, and Symbols*, Part I (New York: The Scarecrow Press, Inc., 1962), p. 223.

30. Joseph Campbell, *The Mythic Image* (Princeton: Princeton University Press, 1974), p. 249.

31. J. S. Calnan and H. J. Copenhagen, "Autotransplantation of Keloid in Man," *British Journal of Surgery* 54/5 (1967): 330.

32. Albert R. McFarland, M.D., quoted in "Discussion," in Bloom, "Heredity of Keloids," p. 519.

33. J.-L. Grignon, G. Scheck, and D. Gross, "La Mauvaise Cicatrice Cutanée: Punition ou destin?" *Revue De Stomatologie* 68/8 (Dec. 1967): 622.

34. Shelley, *Consultations*, p. 284.

35. Omo-Dare, "Yoruban Contributions," p. 372.

36. C. G. Jung, *CW* 9, i, §§576–77.

37. Act II, Scene 1.

TO THE KUGLER/HILLMAN/GOODHEART DRAMA

GREGORY M. VOGT
(Fairway, Kansas)

The Kugler–Hillman–Goodheart dialogue (*Spring 1985*) is the best piece of fiction going now in Jungian thought about practice. We all love a story, a *muthos,* and the more at stake and the more emotion and the higher and more inclusive the philosophical and theological speculation, the better. This story is Dostoyevskyan, in that it involves a patricide and a salvation: it is the renewal of the standoff of Plato and Aristotle—the eidolon versus the metaphysical reality which comes out of the thing itself. It is a wonderful drama with a chorus (the other respondents) who speak the lines the publicum feels but cannot bring to pure speech. It is a story of orthodoxy, betrayal, truth and honesty, tradition, filial piety, sex. There is scarcely an area of human experience or knowledge which is not addressed. Isn't that what brought us to Jung: that we love to talk about everything, and being "Jungian" is a fine excuse for talking about everything, studying everything (hoping to know everything)?

I love this drama because it reminds me of glosses of the Middle Ages, Talmud studies, the careful and yet impassioned explanations San Juan de la Cruz made of his thinly disguised erotic poems to convert them post facto into spiritual exercises. We see contemporary masters at work with the Idea of History, Science (particularly Physics, which gets a lot of airplay today in psychology and esoteric philosophy anyway), The Story of Psychoanalysis, Psychotherapy (what it is and what it is not), Linguistics (in particular, elements of Rhetoric, tropes, metaphor), Logic, Hermeneutics, Theology (special reference, for example, to Goodheart as an "atheist" and what that implies), and, my favorite of all, Ontology and

---

Gregory M. Vogt is a psychotherapist in private practice in Kansas City, Missouri.

a whole series of questions surrounding the status of states and substance in terms of the psyche and in relationships.

Let's look at a few of these areas and the kind of attention and treatment they get. We will be using both philosophical methodology, the kind of speculative glossing that seems to be the "frame" of this dialogue; and looking at the pieces (the Kugler–Hillman dialogue and the Goodheart article that preceded and ignited the former) as they weave together to create a fiction. We will be using elements from Hillman's essay "The Fiction of Case History: A Round" as a key to the second perspective, as well as work by the philosopher William Gass, focusing on the "plot" and "genre" of our story.

A primary complaint about Goodheart's presentation of Jung's dissertation seems to be that Goodheart is "interpreting narrative material as if it were the actual people—Helly and Jung. Worse, he is interpreting their interactions, Jung's motives, and even Helly's psychic health through textual reconstruction" (*Spring 1985*, p. 142). He is seen as being "paradoxical" by treating texts sometimes as " 'autonomous psyches,' while at other times using the very same texts to argue against Jung's concept of the 'autonomous psyche.' Some sections of the dissertation, autobiography, and semi-fictional biography are read by Goodheart as 'absolutely real,' " while others are read as "fictional" (p. 142). Later (p. 143), Kugler has us "notice how his text has subverted the authority of primary texts, including his own." In other words, Goodheart is accused of being unaware of what kind of ontological status he is granting to texts, to people, to the "interactional field" between people. Kugler reminds us (p. 144) that the amount of reality (why not say that instead of "ontological status"—it is not as slick, but it is what is meant) we allow the image in our life will determine how we see the "inner" and "outer" world(s).

There is a lot of attention paid throughout this article to whether Goodheart knows what is real and what is not, whether he can differentiate "real people" from images in a text, whether he understands what is meant by "psyche" or "image," whether he is aware of the dominating metaphors of his language ("frame," "interactional field") and the "ontological" reality he assigns to these. An attempt is made to discredit his work on the basis of this putative lack of awareness and the effect it has on his theoretical propositions.

This game of "courtroom" is too harsh; and Kugler and Hillman's arguments themselves contain too many unjustified assumptions.

Advocating Goodheart's inability to differentiate between text (the written version of what happened) and "actual people" is a strange and very unpsychological interpretation, especially coming from the voice (pen, typewriter, thoughts?—is he "real"?) of James Hillman, who maintains elsewhere that "the entire therapeutic business is a sort of imaginative exercise, and it picks up again the oral tradition of telling stories."[1] Hillman would maintain, as he has throughout his work, that "our existence is imagination."[2] How can one with such a view condemn what could be envisioned as just such an imaginal exercise on the part of Goodheart? One might say (and I imagine Hillman saying) to an analysand who came in with such a condemnation, "yes, well, ontology and such aside, where is that touching *you,* what part of you?" Again, to quote Hillman, "Why does the psyche need to present experience dressed in the costumes of the past as if it were history? Why does the psyche historicize? What does historicizing do for the soul?"[3]

These questions seem much more relevant to psychology than "did this happen?" This seems to be the position Kugler is approaching, but he still maintains an assumption that neither Goodheart nor Hillman is aware of this detachment from ontology; rather, he assumes that they are indeed on two poles of an ontological spectrum, when indeed they might both be either maintaining that in the present discussions there is a final "external reality," or they both may actually be maintaining (and I suspect this) that reality can be "worked" by making certain imaginal assumptions and just going from there to see what comes up.

Hillman was willing to grant that Freud practiced the "historical falsification of memory,"[4] and that "this 'falsification' is nothing other than the historicizing activity of the psyche itself, a history 'making' that is altogether fictional, the poesis of Clio proceeding as an archetypal and autonomous activity."[5] Now that makes Freud really interesting; I wonder why he couldn't do the same with Goodheart's (and Jung's) work. That is to say, he could investigate the claims of Goodheart "psychologically," not literalizing them by arguing their historical and ontological assumptions. And why Kugler, who seems to be tuning in to the literalizing activity of Hillman and Goodheart and exposing both on occasion, still seems to side with Hillman's shifting perspective. I would like to see a dialogue with Goodheart and Kugler now.

Despite how fun and exciting it is, how it stimulates the soul, does the investigation of ontology and ontological assumptions expand our knowledge about the psyche?

From the perspective of the "science of psychology," such investigation is worthless. The formulation of theories in psychology which we hope will add to our science, but which are actually only applications of ontological assumptions, will not help increase our scientific understanding. If we work from the belief that there is an autonomous psyche or a collective unconscious and proceed to demonstrate such, this can never be science. The reason for this, according to Karl Popper, is that there is no possible room for "falsifiability" or "refutation" of the theory. It is based entirely on an untestable hypothesis, hypostatized into a concrete reality. One will have substituted a dogmatism for science; the theory, as we have seen in Jungian, Freudian and Marxist literature to name a few traditions, will be systematically reformulated or widened in order to keep theory and practice in accord. A scientific theory must place itself totally and permanently at risk to be genuine. And even more than Marxism, which at least has some "concrete" assumptions about how the world should be, analysis seems to be utterly unassailable by its impermeability to new data which will shift it truly.

Kugler's example (using it out of context) in the dialogue about the spinning top—that when tapped comes back to its center—demonstrates this centrifugal energy in the theory quite rightly; but it is an unfortunately false analogy between the reality physicists want to demonstrate with this kind of autonomous energy (demonstrable and "physical") and the energy of the "autonomous psyche." The reality or non-reality of the "autonomous psyche" can never be scientifically demonstrated. Experiments which might be performed would involve a reality presumably quite different and with different laws (for example, we assume the psyche to be timeless and spaceless) than a spinning top. Considering Kugler's clear, substantial awareness of the ontological assumptions in Hillman's and Goodheart's arguments, it is strange to see him using a scientific analogy for demonstrating the autonomy of the psyche without a sense of humor and without telling us that he knew he was being metaphorical. Or did I miss the humor?

Hillman and Kugler would be among the first to remind us that Science is a fantasy among others; but again, some demon seems to want them (and Goodheart as well) to literalize here. What is its name?

Is consideration of all these ponderous questions necessary or valuable? Being "enraged" over them? Wouldn't we rather see "all the intractable nineteenth-century problems ... circumvented through a reflexive analysis" (p. 150), as we know could happen with masters such as Kugler and Hillman, with more a sense of wonder and less a sense of the judiciary?

This process is best seen imaginally, as Hillman himself has taught us. Seeing such debates literally—as contributions to Science or History or Philosophy—makes them as unpsychological and deadly (though not without interest in what is unpsychological and deadly) as the case notes of the behaviorist, as Hillman demonstrates in "The Fiction of Case History: A Round."[6] Perhaps what is of more interest to us is not specifically the literal ontologies espoused in the debate as what Hillman calls in the just-mentioned essay the "genre problem." Isn't that what we really want to know, psychologically: what kind of story this is? I think this is a source of the confusion one may sense in the debate about Just What The Hell Is Going On, who's right, who's wrong, who knows what reality is and who doesn't.

What really interests us, I think, is the story. We want to know, is Father (Jung) bankrupt? Why are these guys slugging it out . . . who's winnin'? Will the Soul still exist after this? As William Gass wrote, "The soul, we must remember, is the philosopher's invention, as thrilling a creation as, for instance, Madame Bovary."[7] We want to see this debate as story, starring really not Kugler, Hillman, Goodheart, nor even Jung (who play supporting roles), but the "handsome Absolute." Again quoting Gass from the same essay,

> Although fiction, in the manner of its making, is pure philosophy, no novelist has created a more dashing hero than the handsome Absolute, or conceived more dramatic extrications—the soul's escape from the body, for instance, or the will's from cause.[8]

If it weren't already written so well in the "literature" as it stands—the rage, the betrayals, the misunderstandings, the accusations—one might be tempted to write a "fiction." Goodheart perhaps might be portrayed in our Dostoyevskyan drama as accused of patricide, the bad son, prodigal and wayward, who seemed to claim other fathers entirely despite his patrilineage. He might hold a terrible secret, only part of which he has revealed amidst exaggerations, runs of adjectives, protests of sworn loyalties despite the knife turned against Father: the fool.

Hillman could perhaps play the Good Son, the one who stayed home all these years, learning the craft; harboring resentments, certainly, but on the whole seen as loyal, running the shop (the branch shop) in Father's absence; angered when the quality of the products was questioned. Sure, he gets drunk now and then, gets out of line; says things; redefines the rules. But basically a good Son.

Kugler might be the Rabbi or Priest. He doesn't really want anybody too far out of line. He tells them what they are and are not doing, and makes them see that what they say tells everyone what they are. He's not really Bad nor Good; he is a man of the Word, he is the Servant of the Absolute. It is best, really, to support the son who is running the shop; and perhaps he is a little "under the influence" himself now and then; but (as the rabbi might say), isn't it best we all looked carefully at what we are saying, hem?

But see what I mean . . . what a pale drama this makes in comparison with the one which is already there, written, spoken by these "real and imagined" characters? The fiction as it already stands carries the ambiguities, the tensions and paradoxes which mark good literature. And here the reader may enter as well, as spectator, with huzzahs for his favorite or favorites, with hope that the handsome Absolute may be glimpsed. He—and here I shift back to "I"—may well see this as intrapsychic drama, as a way of seeing the conflicts and assumptions in my own practice, my language "frame," as a kind of "training." We see in the word's history (Latin—to pull, drag, draw) that training is a pulling along. I imagine (in this fiction) myself dragged along by this dialogue, but drawn to it, pulled by it, connected to it, tugging and drawing back; but carried along by "main force." And this love of the story makes me want to say to you: I invite you to look back on this "literature" as if you were reading Joseph Conrad or Flaubert—watch for the dashing entrance of the Handsome Absolute.

1. James Hillman, "The Fiction of Case History: A Round," *in Religion as Story,* ed. James Wiggins (Wolfe City, Texas: University Press of America, 1985).

2. James Hillman, *Re-Visioning Psychology* (New York: Harper & Row, 1977), p. 23. This specific reference reflects a fundamental theme in his work. For a rich treatment of this theme in Hillman's work, see Roberts Aven's *Imagination is Reality* (Spring Publications, 1980).

3. Hillman, "The Fiction of Case History," p. 163.

4. Ibid.

5. Ibid.

6. In "Fiction of Case History."

7. William Gass, *Fiction and the Figures of Life* (Boston: Nonpareil Books, 1971), p. 7. The question of the "autonomy of the psyche" is actually strikingly similar to the question of the "autonomy of Madame Bovary," a theme which recurs in Flaubert's work in the mind of the protagonist as well as in the machinations of the narrator of the novel. She frequently wonders whether she is "real," if she has a "will" of her own. Different literary readers have suggested different resolutions of these issues.

8. Ibid., p. 3.

The Dallas Institute Publications

Lautréamont *Gaston Bachelard*

"A strange, wild book that not only concerns animals, but is itself an animal, to be read through the instincts rather than the mind. The clawing, biting, sucking bounty of the animal psyche leaps out from each page without the impediment of rationality, forcing the reader to confront the brutality of creation. A long needed psychology of transformation that does not sweeten the terrible muscular impulses of the soul, Bachelard's *Lautréamont* opens the eagle's claw, the spider's suckers, the shark's flesh—a bestiary for literary criticism, depth psychology, education, which, to live, must live through animal excess." *Robert J. Sardello*

cloth, 154 pp., indexed $20.00

Water & Dreams: An Essay on the Imagination of Matter *Gaston Bachelard*

"At last this incredibly valuable book in English! No one, not Freud, not Jung, has offered more precise appreciation of the elemental imagination than Gaston Bachelard. For everyone who works with dreams, Bachelard is an indispensable companion."

cloth, 213 pp., indexed $25.00 *James Hillman*

The Bachelard Translation Series

Gaston Bachelard is acclaimed as one of the most significant modern thinkers of France. From 1929 to 1962 he authored twenty-three books addressing his dual concerns, the philosophy of science and the analysis of the imagination of matter. The influence of his thought can be felt in all disciplines of the humanities—art, architecture, literature, poetics, depth psychology, philosophy, language. Forthcoming translations will include *Air & Dreams: An Essay on the Imagination of Movement, The Flame of a Candle, Earth and Reveries of Will: An Essay on the Imagination of Power*, and *Earth and Reveries of Repose: An Essay on the Imagination of Intimacy.*

Stirrings of Culture *Robert J. Sardello & Gail Thomas, eds.*

A collection of concise and provocative essays demonstrating the breadth and courage of vision that has made The Dallas Institute of Humanities and Culture a center of cultural criticism in the brief span of five years. Institute Fellows including Ivan Illich, James Hillman, Jacques Barzun, Wendell Berry, Gail Thomas, Vincent Scully, Louise Cowan, Robert Sardello, Mortimer Adler, Donald Cowan, and William H. Whyte, bring psychological, philosophical, poetic, and practical insight to central concerns of culture in transition—education, architecture, economy, the city, medicine, technology, the body, and imaginative thought.

Publication Date: November 15, 1986 paper, app. 220 pp. $15.00

The Dallas Institute Publications
2719 Routh St. • Dallas, Tx 75201
formerly known as The Pegasus Foundation
DISTRIBUTED BY SPRING PUBLICATIONS • P.O. BOX 222069 • DALLAS, TX 75222

JUNG, THE JEWS, AND HITLER

JAY SHERRY
(Brooklyn, New York)

Since the 1930s the charge of anti-Semitism has been leveled at Jung. The literature is extensive, and though many of the earlier accounts were of a partisan nature a consensus now seems to be forming.[1] It is the aim of this article to review relevant facts and explore new material to produce a more comprehensive, balanced evaluation.

In April 1933, ten weeks after Hitler's assumption of the chancellorship of Germany, Ernst Kretschmer submitted his resignation of the presidency of the General Medical Society for Psychotherapy, a predominantly German organization founded in 1926 to promote the use of psychotherapy by doctors. Since 1930 Jung had been the society's honorary vice-president; Kretschmer's resignation created a dilemma for him. He wrote, "should I, as a prudent neutral, withdraw into security this side of the frontier and wash my hands in innocence, or should I—as I was well aware—risk my skin and expose myself to the inevitable misunderstanding which no one escapes who, from higher necessity, has to make a pact with the existing powers in Germany?"[2] Pleading his interest in protecting psychotherapy, he began to reorganize the society as the International General Medical Society for Psychotherapy. He became its president, introducing rule changes approved at the Bad Nauheim congress in May 1934.

The new society was composed of various national groups, the largest of them German, headed by Martin Göring, a cousin of the Reich Marshal. To avoid German domination, voting was limited to no more than forty percent for any one national group. As an aid to Jewish colleagues

---

Jay Sherry is a teacher, applying the archetypal perspective to the study of history. He has taught at the Institute Rosenberg in St. Gallen while attending the C. G. Jung Institute in Zürich (1977-1982). He now works in the New York City public school system.

expelled from the German section, individual membership was provided for.

The common misunderstanding is that Jung was the president of a Nazified German organization (in fact, Jung actively promoted the formation of new national groups in Scandinavia and later England). This misunderstanding is due largely to developments in the society's journal. Since Jung was president, the society's headquarters were in Switzerland, but the journal continued to be published in Germany. It was agreed that a special edition would be published for circulation only in Germany, from the necessity of all organizations there having to conform to the Nazi ideology. Due to Nazi chicanery the international edition appeared with a Nazi manifesto, catching Jung unawares, embarrassing him by the association of his name with it.[3]

Jung defended his involvement as that of a neutral Swiss lending his reputation to help protect psychotherapy in Germany.[4] He consistently reiterated his nonpolitical stance, maintaining that his interests were professional and he was powerless if a political interpretation were put on his activities.[5] Jung did not foresee that in Europe *everything* would acquire a political interpretation as events began to accelerate.

Besides his activities, we should look at what Jung wrote at this time. In his inaugural editorial in the society's journal he wrote, "The differences which actually exist between Germanic and Jewish psychology which have long been known to every intelligent person are no longer to be glossed over. . . . I should like to state expressly that this implies no depreciation of Semitic psychology. . . ."[6]

The most controversial article Jung wrote was "The State of Psychotherapy Today," which appeared in the journal in 1934.[7] This is significant here for two reasons. First, it contains his most extended analysis of Jewish psychology. Second, his critics tend to focus on one or two passages, ignoring the total context of his remarks. The article opens with an overview of psychotherapy. He criticizes the fact that there are schools which avoid dialogue, hampered by bigoted dogmatism and personal touchiness. Though not mentioning it by name, Jung here refers to psychoanalysis. Dogmatism was one of Jung's standard criticisms of psychoanalysis, and lack of dialogue would refer to the fact that the German Psychoanalytic Society refused to recognize the General Medical Society for Psychotherapy when it was founded.[8]

Stressing personality over technique, he contrasts the theories of Freud (who "took his stand with fanatical one-sidedness on sexuality") and

Adler (who stressed the need for self-assertion). Freud's ideas have exerted influence "largely on the easy opportunity they afford of touching the other fellow on his sore spot, of deflating him and hoisting oneself into a superior position. What a blessed relief it is when one can say in a tight corner, 'That's nothing but resistance,' or when one no longer takes one's opponent's arguments seriously. . . ."[9] "Freud and Adler have beheld very clearly the shadow that accompanies us all. The Jews have this peculiarity in common with women; being physically weaker, they have to aim at the chinks in the armour of their adversary, and thanks to this technique which has been forced on them through the centuries, the Jews themselves are best protected where others are most vulnerable."[10]

There follows the most widely quoted (and criticized) passage:

> As a member of a race with a three-thousand-year-old civilization, the Jew, like the cultured Chinese has a wider area of psychological consciousness than we. Consequently, it is in general less dangerous for the Jew to put a negative value on his unconscious. The "Aryan" unconscious, on the other hand, contains explosive forces and seeds of a future yet to be born, and these may not be devalued as nursery romanticism without psychic danger. . . . The Jew, who is something of a nomad, has never created a cultural form of his own and as far as we can see never will, since all his instincts and talents require a more or less civilized nation to act as host to their development.[11]

Despite the inflated prose the remarks are not in and of themselves anti-Semitic. Comparing the Jews to the Chinese (a culture he greatly admired), he credits them with a wider area of psychological awareness, hardly a derogatory view. In a passage to be quoted later, he speaks of the barbarian component of the Germanic psyche. Such remarks, which appear ambiguous or even contradictory, must be understood with Jung's concept of the psyche in mind. The psyche has a bipolar nature, every phenomenon having a positive and negative aspect. Nowhere in his writings is there a wholly negative view of any individual or group. Jung's therapeutic intention in distinguishing between the psychology of Jew and Aryan was to initiate discussion on a sensitive matter. He failed to appreciate the potential for misunderstanding and misuse of what he was saying. There would be no discussion in Hitler's Germany.

In part Jung's image of the Jew derives from then-current European stereotypes. "The Jew as nomad" was one of the dominant ones (though Jews had lived in Germany for over a thousand years). The Jew is seen as alienated from his chthonic roots—the city-dweller deprived of contact with the soil, the soul-renewing womb of nature. In "Mind and Earth" Jung discusses the influence of the earth on the psyche, observing it is so subtle that one can distinguish among the various types of Russian Jew—the Polish from North Russian and Cossack types.[12] This remark is apparently based upon an anecdote he relates elsewhere, "I know a man who can tell from a series of photographs of Jews of different countries with almost infallible certainty: this is a Polish, that is a Cossack, and that is a German Jew, and so on."[13] Even if we allow such a questionable methodology, a glaring discrepancy stands out of which Jung is unaware. In both passages he refers to a Cossack Jew. No such "type" existed. The Russian Orthodox Cossacks were notoriously anti-Semitic, active in the many pogroms against the Jews in the Ukraine. This should put one on guard regarding Jung's "anecdotal anthropologizing." Psychologically, it is significant that he appears unaware of the anti-Semitic aspect of his characterization and that he knows less about Jews than he thinks.

The 1934 article was not intended as an anti-Semitic tract (as his critics would have it), but as a critique of psychoanalysis. As such it does him no credit.

> In my opinion it has been a grave error in medical psychology up till now to apply Jewish categories—which are not even binding on all Jews—indiscriminately to Germanic and Slavic Christendom. Because of this the most precious secret of the Germanic peoples—their creative and intuitive depth of soul—has been explained as a morass of banal infantilism, while my own warning voice has for decades been suspected of anti-Semitism. This suspicion emanated from Freud.[14]

Outstripping usual criticism of Freud's onesided emphasis on sexuality, he writes that "the main endeavor [of psychoanalysis] seems to lie in trying to explain everything backwards and downwards, and there is nothing that is not capable of some obscene caricature. . . . it merely demonstrates the adolescent smutty-mindedness of the explainer."[15] He sums up psychoanalysis as a "dirty joke" psychology.[16]

In this distorted critique from the former heir apparent of the psychoanalytic movement is a tone that now reminds us of more commonplace critiques. To understand Jung's motives we may take a hint from his repeated emphasis on the personal equation in psychology, which he felt was ultimately a "subjective confession." Jung felt wronged by Freud and his followers, who labeled him a heretic regarding the theory of sexuality, dismissing his psychological approach as mysticism. Once the president of the International Psycho-Analytic Association and editor of its yearbook, he was twenty years later to criticize psychoanalysis from a position as privileged. Stressing its onesided fixation on the infantile roots of neurosis, he promoted his approach as an alternative, one which focused on the neurotic's failure to adapt to the present situation. "We should not try to 'get rid' of a neurosis, but rather experience what it means, what it has to teach, what its purpose is."[17] He writes of medical psychology: "its business is not with neuroses but with human beings . . . and that is why its scope must be widened to reveal to the physician's gaze not just pathological aberrations of a disturbed psychic development, but the creative powers of the psyche labouring at the future, not just a dreary fragment but the meaningful whole."[18]

Jung not only felt wronged by Freud and his followers but also wounded. His comments about psychoanalysts touching the other person's "sore spot" and Jews finding "the chinks in armour" refer not to patients but to "opponents" and adversaries, i.e., Jung. "I am absolutely not an opponent of the Jews even though I am an opponent of Freud's. I criticize him because of his materialistic and intellectualistic and—last but not least—irreligious attitude and not because he is a Jew. In so far as his theory is based in certain aspects on Jewish premises, it is not valid for non-Jews."[19] His sense of woundedness ultimately relates to Jung's transference with Freud, a complex relationship which did not end with their break in 1913.

Jung seemed impervious to the charges of anti-Semitism when they appeared in the 1930s, feeling they had been raised by Freud years before in order to smear his reputation and discredit his work. After he wrote *Symbols of Transformation* (1912),

> Freud suddenly accused me of anti-Semitism. From this I must conclude that I had somehow trespassed against the Jews. This prejudice has stuck to me ever since and has been repeated by all Freudians, thereby confirming every time that psychoanalysis is in fact a Jewish psychology which nobody can criticize without making himself guilty of anti-Semitism. . . . I cannot blame my critics if they succumb to the childish idea that I wrote my books out of sheer resentment. The impassioned delusions of our time impair our judgment. . . .[20]

Parallel to the charges of anti-Semitism are the charges that Jung was a Nazi sympathizer. His first public comment on Hitler was in a radio interview in Berlin in June 1933. He cautioned,

> mass movements have the peculiarity of overpowering the individual by

> mass suggestion and making him unconscious. . . . The self-development of the individual is especially necessary. When the individual is unconscious of himself, the collective movement too lacks a clear sense of purpose. . . . As Hitler said recently, the leader must be able to be alone and must have the courage to go his own way. . . . Every movement culminates organically in a leader, who embodies in his whole being the meaning and purpose of the popular movement. He is an incarnation of the nation's psyche and its mouthpiece. He is the spearhead of the phalanx of the whole people in motion.[21]

Jung was initially impressed by the spirit of nationalism that seemed to be a constructive force. After the war he wrote, "Our judgment would certainly be very different had our information stopped short at 1933 or 1934. At that time, in Germany as well as in Italy, there were not a few things that appeared plausible and seemed to speak in favour of the regime. . . . after the stagnation and decay of the post-war years, the refreshing wind that blew through the two countries was a tempting sign of hope."[22] In those early years of Hitler's rule, Jung was personally affected by those collective forces which were constellated in Germany (he would identify this phenomenon as the archetype of "Wotan" in his 1936 article by that name). Harding relates that Jung

> told us that in 1934 he had gone to Bollingen to work and had put up his yellow flag to warn Professor Fierz that he was not "at home". He was unable to work, however. He felt terribly depressed. A heavy cloud seemed to oppress him. But he kept his flag up and struggled with the oppression all day Sunday and into Monday. At last he pulled down the flag, feeling it was no use trying to work any longer. Immediately, Professor Fierz came over and told him of the Nazi purge, which had taken place on Sunday morning.[23]

As time went on, he became increasingly circumspect in his views. It must be emphasized that his analysis of National Socialism did not mean an endorsement of it.

Besides the "Wotan" article, Jung's analysis of National Socialism is to be found in two interviews given in the late 1930s. Hitler was a mystic medicine man,

> the mirror of every German's unconscious, but of course he mirrors nothing from a non-German. He is the loudspeaker which magnifies the inaudible whispers of the German soul until they can be heard by the German's unconscious ear. . . .
>
> He is the first man to tell every German what he has been thinking and

> feeling all along in his unconscious about German fate, especially since the defeat in the World War, and the one characteristic which colors every German soul is the typically German inferiority complex—the complex of the younger brother, of the one who is always a bit late to the feast. Hitler's power is not political, it is magical. . . . He himself has referred to his Voice. His Voice is nothing other than his own unconscious into which the German people have projected their own selves, that is, the unconscious of seventy-eight million Germans. That is what makes him powerful.[24]

Part of Jung's fascination with events in Germany was due to his conviction that his theory of archetypes was better able to explain what was going on than any alternative theory ("I venture the heretical suggestion that the unfathomable depths of Wotan's character explain more of National Socialism than all three reasonable factors [economic, political and psychological] put together").[25] It is in this light that the widely quoted (and misinterpreted) passage from the 1934 article makes sense. "Has the formidable phenomenon of National Socialism on which the whole world gazes with astonished eyes, taught Freud and his followers better? Where was that unparalleled tension and energy while as yet no National Socialism existed? Deep in the Germanic psyche, in a pit that is anything but a garbage-bin of unrealizable infantile wishes and unresolved family resentments."[26] Jung intended this as a vindication of his theory, not of Hitler.

Challenging psychologists to consider their cultural/ethnic assumptions, Jung wrote, "I am also quite ready to suppose that I am a bigoted Swiss in every respect. I am perfectly content to let my psychological confession, my so-called 'theories' be criticized as a product of Swiss woodenheadedness or queer-headedness. . . ."[27] Jung was a conservative, upper-class Swiss. Surveying the European political landscape, he felt the best policy toward Germany was to maintain a "give-it-a-chance" attitude (as exemplified by Chamberlain). In his 1938 interview he said, "the only way to save Democracy in the West—and by the West I mean America too—is not to try to stop Hitler. . . . you can only hope to influence the direction of his expansion. . . . Let him go to Russia. That is the logical *cure* for Hitler. . . . Nobody has ever bitten into Russia without regretting it. . . . It wouldn't matter to Russia if somebody took a bite. . . ."[28] This seemed a neat solution—let the two totalitarian powers battle each other. This suggestion became reality, with the most appalling results (twenty million Russians dead, thousands of towns and villages destroyed).

If Jung was neither anti-Semite nor Nazi sympathizer, what did he mean when he admitted to Rabbi Leo Baeck after the war that he had "slipped up"?[29] In his postwar article "After the Catastrophe" he wrote,

> no article has ever given me so much trouble, from a moral as well as a human point of view. I had not realized how much I myself was affected. There are others, I am sure, who will share this feeling with me. This inner identity or *participation mystique* with events in Germany has caused me to experience afresh how painfully wide is the scope of the psychological concept of *collective guilt.* So when I approach this problem it is certainly not with any feelings of cold-blooded superiority but rather with an avowed sense of inferiority.[30]

What prompted this humbling admission? In spite of often brilliant analyses, what had been missing in his prewar viewpoint? Like the majority of people at the time, he failed to appreciate the depth of *evil* incarnate in Hitler and his movement, in spite of the fact that as early as 1918 he had seen the "blond beast" stirring in the German psyche. In a postwar letter he wrote, "before the Hitler era I still had some illusions which have been radically destroyed by the prodigious efforts of the Germans. I really had not thought that man could be so absolutely bad. I had thought he could be evil, but evil has at least a certain character, while evil in Germany was rotten."[31] The medical and political neutrality he claimed in the 1930s was unfortunately accompanied by a *moral* neutrality which helps explain the ambiguous tone of his contemporary comments. In "After the Catastrophe" he continues his lifelong involvement with Faust: "we never get the impression that he has real insight or suffers genuine remorse. His avowed and unavowed worship of success stands in the way of any moral reflection throughout, obscuring the ethical conflict, so that Faust's moral personality remains misty."[32]

Any complete account of Jung must consider what unconscious material is available. In this case the most significant is a dream he had in August or September 1939, related after the war to Esther Harding. I will give the dream and Jung's interpretation in full.

> He found himself in a castle, all the walls and buildings of which were made of trinitrotoluene [dynamite]. Hitler came in and was treated as divine. Hitler stood on a mound as for a review. C. G. was placed on a corresponding mound. Then the parade ground began to fill with buffalo or yak steers, which crowded into the enclosed space from one end. The herd was filled with nervous tension and moved about restlessly. Then he saw that one cow was alone, apparently sick. Hitler was concerned about this cow and asked

> C. G. what he thought of it. C. G. Jung said, ''It is obviously sick''. At this point, Cossacks rode in at the back and began to drive the herd off. He awoke and felt ''It is all right''.

His interpretation:

> Hitler was treated as *divine*. Consequently, he felt, we had to view him like that, that Hitler is not to be taken primarily as a human man, but as an instrument of ''divine'' forces, as Judas, or still better, as the AntiChrist must be. That the castle was built of trinitrotoluene meant that it would blow up and be destroyed because of its explosive quality. The herds are the instincts, the primitive, pre-human forces let loose in the German unconscious. They are not even domestic cattle, but buffalo or yaks, very primitive indeed. They are all male, as in the Nazi ideology: all the values of relationship, of the person or individual, are completely repressed; the feminine is sick unto death, and so we get a sick cow. Hitler turns to C. G. for advice, but he limits his comment to the diagnosis ''the cow is very sick''. At this, as though the recognition of the ailment released something, the Cossacks burst in. Even before that, the herd had been disturbed and nervous, as indeed the male animal is if separated too long or too completely from its complement, the female. The Cossacks are, of course, Russians. From what C. G. said, he deduced that Russia—more barbaric than Germany, but also more directly primitive, and therefore of sounder instinct—would break in and cause the overthrow of Germany.[33]

Jung had this dream after the Hitler-Stalin pact was signed on August 23. This non-aggression pact stunned the world—the ideological arch-enemies had put aside their differences, silenced their propaganda machines, and reached an agreement that sealed the fate of Poland. It was a time of great anxiety, and Jung said it looked ''as if they would be swallowed up without a doubt.''[34]

Jung's interpretation is both incomplete and unsatisfactory. One of his great contributions to dream analysis is his distinction between objective and subjective interpretation. An objective one relates dream figures to actual people in the dreamer's life; the subjective takes the figures as personifications of different aspects of the dreamer's personality. Although both are important, Jungian dream analysis tends to emphasize subjective interpretation. In this one, Jung is exclusively objective, turning the dream into another example of his prescience regarding historical developments in Europe (reminiscent of his pre-World War I vision of a European bloodbath). We must pursue a subjective interpretation to reach a deeper understanding of Jung's unconscious situation.

"In Hitler, every German should have seen his own shadow, his own worst danger. . . . He was the most prodigious personification of all human inferiorities."[35] In the dream Jung is on intimate terms with the shadow and is partly identified with it (raised up on a similar mound).

The personality is often symbolized by a house. That it is here a castle relates to Jung's medieval interests. Ominously, it is constructed of TNT. We usually associate castles with security for beings inside; here its explosive nature would pose the greatest danger to those within. Into this explosive psychic space crowds the herd—collective instincts. They are in a state of disharmony.

For the Cossacks we must connect the dream image with Jung's conscious attitude. He had a low opinion of the Russians ("more barbaric than Germany," "the Slav mujiks, the poor white trash of Europe"[36]) and had recently advocated a German invasion of Russia, so the appearance of the Cossacks forebodes revenge, not liberation, an attitude we also find in his "Wotan" analysis. Stressing again and again that Wotan is rooted in the German psyche, he writes, "The disturber of the peace is a wind that blows into Europe from Asia's vastness, sweeping in on a wide front from Thrace to the Baltic, scattering the nations before it like dry leaves. . . ."[37] This analysis shifts the source of aggression to Asia (Russia). The dream took place at a time of great anxiety when Jung feared being "swallowed up" (a reversal of his comment that "it wouldn't matter to Russia if somebody took a bite").

His waking up from the dream saying "it's all right" might be considered a reaction formation to help suppress the anxiety generated by the dream. Dreams "break through when their function demands it, that is, when the compensatory contents are so intense that they are able to counteract sleep. A compensatory content is especially intense when it has a vital significance for conscious orientation."[38]

The key to the dream is Jung's diagnosis of the cow. It is the point in the dream that he calls the culmination—the point at which something happens which determines the outcome of the dream. The scene is dominated by a herd of male cattle which Jung associates with the overly masculine Nazi ideology. The solitary cow is sick. This deeply divine image connects us with the nurturing mother who gives life-promoting milk. This fertile, bovine aspect of the anima is in a dangerous situation. What made the cow so sick? Jung's fascination with the Wotan archetype alive in Germany and his bitterness toward his old psychoanalytic colleagues caused him to miss out on a feeling level what was happening there.

''The more onesided his conscious attitude is, and the further it deviates from the optimum, the greater becomes the possibility that vivid dreams with a strongly contrasting but purposive content will appear as expression of the self-regulation.''[39] Not only dreams but physical symptoms may appear. There is evidence that during the war years this anima problem both continued and moved toward resolution. In December 1941 he wrote, ''I begin to feel my age and whenever I get a bit too tired I also feel my heart. . . . I went through a period of black depression during this first four days.''[40] His correspondent is an American; the reference is to the Japanese attack on Pearl Harbor on December 7. An unconscious component of the depression might be related to the Russian counterattack which began on December 6. The Germans attacked Russia on June 21 of that year, scored stunning successes, and threatened Moscow. Relying on General Winter and its vast reservoir of Siberian manpower, the Russians took the offensive. After the defense of Stalingrad, the Russians pushed the Germans out of Russia and swept into Europe ''on a wide front from Thrace [Bulgaria] to the Baltic.'' Like his depression in 1934, this one may have resulted from his unconscious identification with the ''German spirit.''

In 1944 Jung broke his foot and shortly afterward suffered a heart attack. We should take a closer look at the role of the psyche in these somatic events. Jung once wrote that ''the moral reaction is the outcome of an autonomous dynamism fittingly called (among other associations) the heart.''[41] Perhaps Jung was experiencing something new, a profound moral reaction to the war, rediscovering the ''sick cow'' of his moral neutrality of 1939.

At death's door he experienced a series of visions which he describes in his autobiography *Memories, Dreams, Reflections.*[42] In the hospital at his evening meal, he saw his nurse as an old Jewish woman who served him kosher food. This is an important development: from a sick anima to a maternal anima figure who nourishes him from the tradition of a people he had not sufficiently appreciated. This was followed by a series of visions of the *mysterium coniunctionis* in one of which he was Rabbi Simon ben Jochai. He is now identifying himself with a Jew, joining the anima in a sacred marriage, symbol of wholeness, thus transcending the claptrap about ''Jewish categories,'' reconnecting himself with that universal humanity at the heart of the collective unconscious.[43]

These visions indicate the change of heart taking place in Jung, made conscious and public by his postwar admission that he had ''slipped up'' and his discussion of collective guilt. Continued reflection on the moral and psychological issues culminated in his *Answer to Job.*[44]

1. A typical psychoanalytic account can be found in F. Alexander and S. Selesnick, *A History of Psychiatry* (New York: Harper & Row, 1961), Appendix B, pp. 407–09. The most comprehensive Jungian account is A. Jaffé, "C. G. Jung and National Socialism," in *Jung's Last Years and Other Essays* (Dallas: Spring Publications, 1984). More recently, Grossman, "C. G. Jung and National Socialism," *Journal of European Studies* 9 (Dec. 1979): 231–59.
2. C. G. Jung, "A Rejoinder to Dr. Bally," *CW* 10, §1016.
3. C. G. Jung, *Letters,* vol. 1 (Princeton: Princeton University Press, 1973), pp. 144–45 (March 2, 1934).
4. See J. Cocks, *Psychotherapy in the Third Reich* (New York and Oxford: Oxford University Press, 1985). Cocks shows that psychotherapy was able to maintain a surprising degree of independence under the Nazis. This was due, however, to factors other than Jung's influence.
5. C. G. Jung, *CW* 10, §1034.
6. *CW* 10, §1014.
7. *CW* 10, §§333–70.
8. Cocks, *Third Reich,* p. 6.
9. *CW* 10, §347.
10. *CW* 10, §353. Also, "The Role of the Unconscious," in *CW* 10, §19.
11. *CW* 10, §353.
12. *CW* 10, §93.
13. *CW* 10, §968.
14. *CW* 10, §354.
15. *CW* 10, §355.
16. *CW* 10, §356.
17. *CW* 10, §361.
18. *CW* 10, §354.
19. Jung, *Letters,* vol. 1, p. 154 (March 26, 1934).
20. Ibid., pp. 167–68 (March 19, 1934).
21. *C. G. Jung Speaking,* ed. McGuire and Hall (London: Thames and Hudson, 1978), pp. 63, 64, 65.
22. *CW* 10, §420.
23. *Jung Speaking,* pp. 183–84.
24. Ibid., pp. 118, 119–20.
25. *CW* 10, §385.
26. *CW* 10, §354.
27. *CW* 10, §1026.
28. *Jung Speaking,* pp. 132, 133.
29. Jaffé, *Jung's Last Years,* pp. 97–98.
30. *CW* 10, §402.
31. *Letters,* p. 425 (April 20, 1946).
32. *CW* 10, §423.
33. *Jung Speaking,* pp. 181–82.
34. Ibid., pp. 180–81.
35. *CW* 10, §454.
36. *Jung Speaking,* p. 123.

37. *CW* 10, §391.

38. *CW* 8, §487.

39. *CW* 8, §488.

40. Jung, *Letters,* p. 307 (Dec. 20, 1941).

41. *CW* 10, §843.

42. C.G. Jung, *Memories, Dreams, Reflections* (New York: Pantheon Books, 1962), p. 294.

43. This identification with a Jew also appears in his postwar article "After the Catastrophe." He writes that he felt like the prophet at the siege of Jerusalem whose lamentations were of no avail and who was killed by a stone missile—*CW* 10, §414.

44. See H. Slochower, "Freud as Yaweh in Jung's *Answer to Job,*" *American Imago* (Spring 1981): 3–39; J. Groesbeck, "A Jungian Answer to: 'Yaweh as Freud'," *American Imago* (Fall 1982): 239–53; J. Kirsch, "Reconsidering Jung's So-Called Anti-Semitism," *Arms of the Windmill* (Baltimore: Lucas Pub. Co., 1983), pp. 5–25.

The Master's Culinary Memoirs

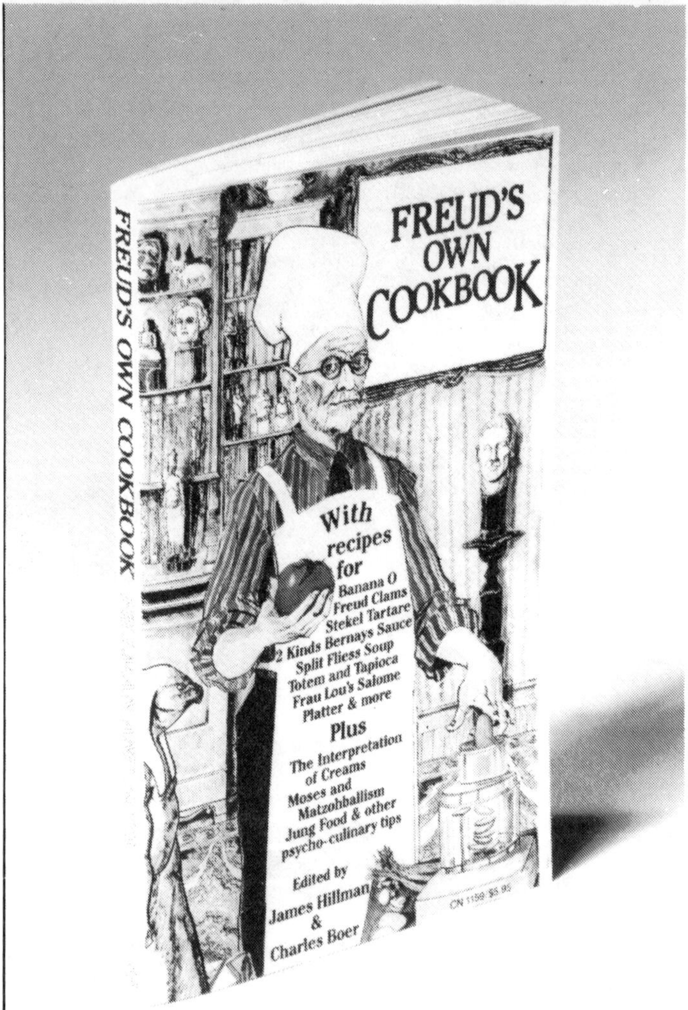

Find out about
Civilization and
Its Indigestion,
Luncheon Interruptus,
and The Interpretation
of Creams.
Try actual recipes for
Freud Clams,
Krafft-Ebing Lady's
Slippers, and
Slips of the Tongue
with Sauce Bernays.
Sigmund Freud,
that irrepressible
master chef,
has the last word:
**"What do women
want?
They want lunch!"**

Edited by JAMES HILLMAN and CHARLES BOER

In paperback, at bookstores or call **TOLL FREE (800) 638-3030**

Harper & Row 

DOMESTIC LIFE WITH C. G. JUNG
Taperecorded Conversations with Ruth Bailey

GLIN BENNET
(Bristol, England)

Ruth Bailey had been a friend of the Jung family and a regular visitor to the house at Küsnacht for about thirty years. She was an unusual friend because, though admiring C. G. Jung tremendously, she was never obviously in awe of him. She treated Jung in her warm, direct north-country English way which he liked, and which had a relaxing effect on the austere Jung household, unaccustomed to laughter and jokes. So she was always a welcome guest there, all the more so because she had no particular interest in psychology, Jung's or anyone else's, and no overt romantic relationship with him, so she did not make emotional demands or pose any kind of threat to the family network. In time she would become an integral part of it.

After Emma Jung's death in 1955 the family had no idea how they were going to look after Jung, but it was Jung who took the initiative. He took the unusual step for him of telephoning Ruth at her home in Cheshire and asking her to come and look after him for a while. This temporary arrangement was to last for the rest of his life and was to bring him great contentment. There were of course various ladies in the vicinity who would have been glad to move in and care for him, but they were all psychologists or otherwise connected with the analytic scene, and some would have made demands on him that he no longer wished to meet.

Ruth Bailey was born in 1900 and so was twenty-five years younger

---

Glin Bennet is a Consultant Senior Lecturer in the Department of Mental Health, University of Bristol, a Consultant Psychiatrist, and a psychotherapist. He is author of *Patients and Their Doctors: The Journey through Medical Care* (London: Bailliere Tindall, 1979), *Beyond Endurance: Survival at the Extremes* (London: Secker and Warburg; New York: St. Martin's Press, 1983), and other works on medical and psychological topics.

than Jung. They met in Kenya which was the starting point for his East African travels in 1925.

Ruth was traveling to Kenya for her sister's wedding, and though she had probably met Jung and his party on the boat, they did not meet properly until they reached their hotel in Mombasa. As she put it, "I am not sure whether I picked CG up or he picked me up."

It happened during a fancy-dress party in the hotel. Jung was actually out in the lounge studying the maps for his journey to Mount Elgon when Ruth came over and took an interest in what he was doing. He asked her if she liked maps. She said she did, and after a very brief conversation it seems Jung had invited her to join his expedition, which included Godwin Baynes and George Beckwith, an American friend. She accepted at once despite the protests of her family and the serious misgivings of the other members of the expedition.

It was quite a masterstroke of Jung's to incorporate Ruth into this odd male group, especially as Godwin Baynes's first wife had just committed suicide and he wanted to lean extra heavily on Jung, and George Beckwith was an unknown quantity. They experienced the usual delays and vicissitudes associated with travel with porters in the African bush, and Beckwith developed malaria. The Africans they encountered seemed to recognize Jung as some kind of Great White Magician, perhaps through a combination of his massive frame and his ability to empathize with them, and that smoothed their passage. Even so it was fortunate to have Ruth along, as she was a nurse and could look after these men and make a joke when the atmosphere became too tense, as it frequently did.

On their return to Europe, Ruth went to stay with the Jungs at Kusnacht, where she told Emma all the small details of the trip and the absurdities which Jung himself did not see fit to relate. Emma liked Ruth and the way she kept CG happy without there being any emotional entanglement, which in 1925 was a sensitive issue because of Jung's involvement with Toni Wolff.

*

I do not think anyone could have been the perfect companion to Jung. In his middle years he had a highly intelligent, well-grounded woman, Emma, as his wife and the mother to his five children; and he had Toni to nurture him and help him with his work, and to follow him in his psychological explorations and his own confrontations with the unconscious. Then towards the end of his life, when I imagine he was to a great extent emotionally self-contained, there was Ruth Bailey, a lifelong friend who knew and tolerated his little ways, though at times he may have been slightly irritated by her stories and jokes. However, he recognized her worth and her devotion to him, and eventually he became quite dependent on her. After Jung died in June 1961, Ruth Bailey returned to her home in Cheshire where she lived until her death in 1981. I used to visit her quite often and was fascinated by her stories about life at the Jungs'. She agreed to record some of these stories and for me to make what use of them I wished. Her record is very much domestic chat and makes no claim to be a contribution to Jungian thought. Nevertheless, it gives another view of the man in his later years, always in danger of disappearing into an image of a benign white-haired contemplative who carved in stone.

Interview

Ruth Bailey starts by relating a story of herself and Jung sitting at Port Said at the end of their long journey through Africa which had begun several months before at the hotel in Mombasa.

Ruth Bailey: We were killing time, really, and we sat there and said nothing, and suddenly I said,

> The sea no father has, nor loving mother,
> No little sister fair, or loving brother,
> And that is why when ill at ease
> You sometimes hear the sobbing sea.

And then CG said, "Really, Ruth, your brain is like a brawn, little bits hung together with jelly."

"That may be, but it entertains you."

"Yes, you are quite right, it does."

I don't think he ever met one before like that. I had quite a lot of odd bits of verse that I remembered that I have collected in my lifetime, and they always entertained him. He never knew what was coming next. . . . And he loved limericks. I made one up about him. A cardinal came from Rome to see him to really say that one of his books ought not to be published, but it should only be used by the Jesuits and teachers. When the cardinal went away I said,

> There was once a doctor named Jung
> Who to his theories gave tongue.

The Church was afraid of the noises he made
And the very strange songs that he sung.

He was so pleased with this that I had to write it down in his own book.

Once, when he was going to Schaffhausen for two days (I was staying there as a visitor in the summer. I was a guest and it was long before I went to live with them because Emma was still there, too), he said he had had a dream, a very funny dream, the night before. He was a beggar in rags outside rather a palatial sort of place with pillars, and he went in and out of the pillars but his begging bowl was gold. And I said, "Well, that's in the Bible, CG."

"Where is it in the Bible?"

"Well, I wouldn't know, but it says 'Never the golden bowl be broken or the silver cord be loosed.' I know that much because I loved these words, you see. I love words, I do, they conjure up pictures."

And he said, "You find it for me."

"I wouldn't know where to look."

"Well, while I am in Schaffhausen, go into my library and sit at my desk and have the Bible there and find it for me."

"Where shall I look?"

"Try Numbers, Ecclesiastes."

And myself those two days going down the columns and up the columns, and found it—Ecclesiastes, chapter 12. When he came back I said, "There you are, I found it."

And he said, "Yes, you're like my violin. I can play on you and get the tune." Now, wasn't that nice.

Glin Bennet: I felt at a personal level you were closer to him than anyone.

Ruth Bailey: Yes, he said so. I have a letter where he said so. He said that I had kept him alive. But I think it was so because it was something out of the ordinary for him. You see, he had all these clever women, and some of them were very clever. Look at [Jolande] Jacobi and all those people, they were very clever and they could talk to him and discuss things with him, but it was all so inhuman somehow. . . . They drained him, because he would be exhausted at the end of an hour with some of them.

The greatest thing I could do was to make him laugh. Make him see the funny side of something, which he could see—he'd got a great sense of humor. They say the Swiss haven't, but he had.

GB: Who could he take his troubles to?

RB: Nobody, because he couldn't take them to his family, you see—only Aggie [his eldest child]. Aggie was the one. He never had any worries about Aggie. She stood on her own two feet, and she followed her instinct, which was quite right. But the others—of course all of them were very stiff with him when I first went there. They shook hands with him, you know, they never kissed him goodbye or kissed him when they came. And Marianne would say, "Goodbye, Father." I would say, "Oh, Marianne, for heaven's sake, kiss him, couldn't you?" And he looked at me, and his eyes twinkled. Then she did afterwards.

They were rather afraid of him. That was the trouble—they had been brought up to be rather afraid of him, and I bet he was rather a frightening person in those days, middle age, because he was coming up against tremendous problems of his own. And you know, he nearly shot himself once.

He tells you about it in *Memories, Dreams, Reflections*. It was before we went to Africa. . . .

GB: . . . during the time of his confrontation. . .

RB: . . . with himself, and also he was so ostracized. When he started up on his own in Küsnacht, they used to nudge each other in the train and say, "That is where that old wizard lives."

GB: Toni Wolff was a support to him then.

RB: Yes, tremendous, she was. Right up till alchemy and religion. She couldn't take alchemy. She let him down very badly with that, and she wouldn't go along with him, and I think he was terribly disappointed. She couldn't take it somehow. She wouldn't accept what he said. And, of course, in a way he wanted bolstering up with it, not saying, "I don't believe it." And also another thing, I think then she started being very ill with arthritis, because her fingers went like sausages. She was in an awful state and used to have this gold stuff, and I think in the end she had too much of it. That's really what killed her. But she used to come to Droitwich [Spa] for the treatment.

[Toni] was there when he needed somebody, because Mrs. Jung was so like a Swiss woman, completely taken up with the children, and, of course, there were five of them and the household, and she had been so conventionally brought up. If you saw her home in Schaffhausen and the old lady [Emma Jung's mother] and the old maids they had. When I went there with her, the old maids were still there that they had had all their lives, in white caps and streamers, and beautiful starched aprons. They'd been there all their lives.

The old lady [would say]: "You sit next to me, Miss Bailey, and talk English to me. Tell me about England." She was a darling, and of course she loved CG. She took him for walks when he was a little boy of two [more likely when he was four], and she was a young woman. She never knew then that he was going to be her son-in-law.

I think Miss Wolff came in then as a very intelligent woman and, of course, adored him, like they all did, and bolstered him up through very difficult years when he was in a way cut off from his family and cut off from Emma, you see. . . .

He never told me much about [his relationship with Toni Wolff]. Towards the end, she used to come out to Küsnacht on Wednesdays for tea, and to begin with they used to have it in the library, the two of them, and then when it was summer and I was there, and Emma was in hospital, he had it in the garden with her. I'd got the tea ready and told CG to bring it out to us, and I'd always remember this occasion. I said, "Right, I think it is all here, CG."

And he said, "You don't leave us. You stay here."

And I thought, "Oh, my goodness." I didn't know what to do.

He said, "You pour out for us."

And then when we finished tea and we were talking, he got out his little paperback book and started to read it, and didn't take any notice of her. I felt ever so sorry for her. I thought that was awful, really. Terribly bad-mannered of him. I told him so after she had gone.

"Do you think so?" he said.

"Yes," I said, "It was very bad manners."

Anyway, I ended up feeling quite sorry for her. She would come, you know, looking beautifully turned out and everything, but it had gone. Whatever there was between them had gone completely. There just wasn't anything.

To begin with, Toni had no use for me whatsoever. When I first turned up from Africa, I was very much in the dog-coat, so to speak. She simply ignored my existence, but I didn't care. CG did talk in Africa about her; that was the only time. He talked with Dr. Baynes about her and about her intellect, her mind. That's the first I heard about her, and then Beckwith told me a bit about her because he knew her. He said she was terribly clever and was CG's assistant, he called her. But I never heard anything really much more. The rest of it I gathered when I stayed with them.

CG and I were playing cards, Patience, I think. I can see us now in that big salon, and just as the door slid back and she [Toni Wolff] came through—by this time she was quite friendly with me, better anyway—just as it slid back she heard me say, "You are an old cheat, CG. You cheat. You wouldn't play with us in England, we wouldn't have you cheating like that."

And the door was hastily shut, and she disappeared. To say this to the God was terrible, wasn't it? He knew he cheated, too. He did cheat on purpose. But it was so funny. It amused me, these reactions, because apparently I blundered in where angels feared to tread because out it came, and, of course, that was what he liked—he roared with laughter.

GB: How did Emma and Toni get along together?

RB: On the surface they had perfect manners; this was towards the end, of course. I didn't know earlier on because I didn't see them together. But Toni was President of the Club, and Emma would sit there on the council of the Curatorium under her, and then in the end Emma became President. Emma of course had perfect manners, and if Toni was at the house she was treated as a guest and so forth. I never knew Emma's reaction to it, except I always thought she looked a bit sad. But towards the end of her life she came into it all right. She got through.

*

Ruth Bailey spent a good deal of time with Jung at Bollingen. Life was basic there, and much of their time was spent on domestic activities. I asked her about his feelings for things around him and how it manifested itself in daily life with him.

RB: He loved wood, and he loved messing with it, you know, chopping it and doing something. And we used to cross-cut great chunks together, and he'd quote me the thing from the Oxyrhyncus Papyrus.

I said, "CG, would you do something for me. Would you please copy out those verses you say to me sometimes?"

And he did; he copied it out for me:

> Wherever there are two they are not without God.
> And wherever there is one alone, I say I am with him.
> Raise the stone and there thou shalt find me,
> Cleave the wood and there am I.

I think that's marvelous. [She relates further words Jung often quoted.] "The greatest of all lessons is to know oneself." Jesus said, "Let not him who seeks cease until he finds. And when he finds he shall be astonished.

And astonished he shall reach the Kingdom. Having reached the Kingdom, he shall rest.''

We were [sailing] on the lake for the day. We used to take a basket of fruit with us. We'd got quite far up and it darkened very suddenly, and he thought we'd better turn back as there was a storm coming and the yellow flags were going up. We turned round and we came like the wind. I don't know how he held the tiller and the sail. I clung on to the sail. We just came back like an express train. It had taken hours to get up there, and we came back in a few minutes.

Emma would never go with us. She didn't like sailing. She used to stay home in Bollingen and do her work and her lessons, like her reading and everything, and said, ''You go with him, Ruth.'' I was terrified sometimes. And I had to learn about all these ropes, you see, which ones to pull, and I always pulled the wrong one! He cursed me, and I said, ''God, I won't be your cabin boy anymore if you do that; you should have patience with me.''

I got into it quite well in the end.

CG did all the cooking. We prepared it, Emma and I did, and she did exactly what he wanted and so forth—and got herself into trouble for doing it. He'd gone up to his study to write, and he wanted cauliflower, I think, for supper. We always had heaps of vegetables sent in baskets from Küsnacht, and this cauliflower was bad, and instead of throwing it away, she went up to the study and showed it to him. Of course he went off the deep end and said how terribly wasteful that was, and went for her, you see. She came down in tears. I said, ''Emma, why ever did you go and show him that cauliflower? I wouldn't have done it.''

''What would you have done?''

''I should have thrown it away and given him something else.''

''Oh, but he said cauliflower.''

''Well, he wouldn't have remembered. There's such a lot of vegetables in the cellar. Say you got this one instead. Or, did he say cauliflower? Well, you got this one. I should have put it in the hole,'' where we threw everything.

''Oh, I couldn't do that, not when he had ordered it.''

''Of course you could.'' I used to talk to her like that. ''It isn't worth getting upset for, and he's a pig to go on to you about it.''

Of course, that was heresy, too. Anyway, when it came to my turn, I got the better of it, though. When it came to my turn to do the Bollingen preparing and everything else, there was no me going to him and telling him something was bad; it had gone in the hole a long time ago if it was going to be bad. I worked it out this way.

"You see, Emma," I said. "You went up and showed him this cauliflower. You really expected and waited to be told off. He couldn't do anything else. He couldn't say, 'Well, never mind, my dear, give us a kiss.' You just asked for it and you got it." Then I made her laugh.

I got it myself finally. A long time afterwards, a long time after Emma had gone and he let me actually do the cooking then. He was writing and so forth. I was making sougetti, made with vegetable marrow cut up with meat in a sort of stew, and I put two tomatoes in it, as Marianne did. Well, he was furious with me, absolutely furious. He'd never told me to put tomatoes in it, and I had spoilt it and he'd never trust me again. Then I was angry and I said, "Well, if your trust is only worth two tomatoes, I don't think much of it. Really, CG, that's a bit off."

He was still furious and wouldn't have any dinner, and I was awfully upset and I said, "Well, of course, it's about time I went. You had better try and find somebody else to look after you and I'll leave you tomorrow, because if you cannot trust me, it's no use me staying, is there."

I had a miserable night thinking about all this. But he was really very unreasonable about it, for, dash it, what was two tomatoes to fuss over!

So, in the morning he had breakfast as usual—very cold and taciturn. He said, "Now leave the things, Ruth. I want you to walk with me beside the lake."

So I said, "Yes, I will."

"I want to ask you not to leave me today."

"Of course, I won't, CG. I never intended to."

Then a twinkle came in his eye. "Oh," he said, "I was afraid you did."

"Well, for a moment I did because you were so very unjust about those two wretched tomatoes. You know, if you had been a nice well-mannered Englishman, you wouldn't have mentioned tomatoes." And then he did laugh. "You should have ignored them. It wasn't such a crime as all that."

"Oh well," he said, "Never mind, but I don't like tomatoes, anyway."

"I know, I realize that now, and I won't ever do it again, but really, you were a bit too bad with me, weren't you?"

"You must never forget," he said, "that I once had red hair, and I've got a frightful temper. Don't ever rouse my temper, Ruth."

"Well, I certainly won't do it again if I can help it."

But he said, "What I want you to do is to see me out, will you? Will you stay with me and see me out?" And that's when he asked me to stay.

"Yes, I think I could do that," I said, "but I must remember about the tomatoes, mustn't I."

And, you know, he loved tomatoes—he loved them, but they always had to be out of the garden, he couldn't bear the bought ones.

In the kitchen [at Bollingen] every single thing had its proper place, and if one spoon was out of its proper place, he knew. He was furious if they were in the wrong place. Everything had to be just so. And, of course, that is the sailor's way, too. But his carving tools and everything hadn't to be interfered with at all, and it was all mapped out, and chopping the wood. We even chopped the twigs.

GB: Could he always work, or were there times when he just couldn't?

RB: He didn't sit still doing nothing very much. Perhaps a bit at Bollingen after a meal, if it was nice, he would sit in a chair with his pipe or a cigar looking over the lake, but he didn't waste much time. Then, of course, if there was a bit of time, he walked along to the boat place, along the lake, and walked around the estate, so to speak. I told you about the tom-tit, didn't I. He was sitting in the garden by himself. He would often sit there in his chair reading something—an essay or a lecture. He was very still, and the tom-tit came on his head and tried to pull a hair out. I think that was lovely. Those little, tiny feet clung into his head. The first he knew about it was those little feet like pins into his head—and with those long silvery hairs, the bird evidently thought that this was a marvelous thing for its nest.

And he used to watch the fishes. There is a lovely picture of him watching the fishes. We made porridge in an iron pot, and it stayed in that pot for three weeks with more being added to it until it was really a lovely cream color. Of course it was boiled every day. But when it had to be changed, we never tried to wash the pot; we put it in the lake and the fishes did the rest. You could see them there; they had a wonderful time, and they took every scrap off. It would have taken us hours to clean. And then we cleaned it after that, but it was yellow inside. They had a marvelous time, and he'd sit there watching them do this. He loved everything like that.

He could be up in his study miles away and could come down to earth again. I said to him one day, "CG, you know everything."

"I don't, you know. But, you see, I just manage to keep balance, and when I go to Bollingen, it balances again. I go over the other side."

And his being close to the earth, loving all things earthy, kept him balanced. This tremendous thing on this side—intellect—and then his love of all the earth of this side. Really, it was marvelous to watch him weeding, sitting on a little stool weeding, and planting things—he planted potatoes and corn. He was a real peasant at Bollingen. And the peasants there absolutely loved him. He let two fields to a peasant in Bollingen—a peasant farmer there who had about three sons. They absolutely adored him. He always used to go and talk to them when they were making hay. And he used to walk about with a long stick that he had carved, like a shepherd's crook almost.

GB: Going on now to the last years, was there a time when you felt his life was drawing to a close?

RB: Well, the first things were the dreams he told me. I told you of the dream he had of this big, enormous round stone in a high place and on it was inscribed, "And this shall be a sign unto you of Wholeness and Oneness."

And he said, "You see, Ruth, I am not going to be here long."

"Don't say that, CG. Don't say that. Whatever shall I do when you go and leave me?"

"Well, I'll do my very best to welcome you when you come the other side."

It was nice. And then we didn't talk about it anymore. And then he had another dream. That was extraordinary. It was as if a lot of trees made a square, all fibrous roots, coming up from the ground and surrounding him. There were gold threads among the roots.[1]

GB: It must have been a strange moment when he told you about the dream of the stone.

RB: Yes, but that was only about three weeks before he died.

GB: What did he feel then?

RB: He didn't mind. He thought it was finished. He was thinking from the time when he was so very ill [the summer before] that it was drawing to a close. He couldn't do the same [again], and he had to be very extra careful. He couldn't do any more trips out with Fowler [McCormick, an old friend], and I think that upset him very much because he loved driving out with Fowler.

GB: Did he feel that he had completed his work?

RB: Yes, I think he did. He couldn't do any more, he said. And he had completed *Dreams* and everything else, and the last two chapters were in English—didn't write in German.[2]

GB: His last writing . . .

RB: Was in English. Yes, and he told me that it was easier for him to write in English than it was in German because "I can get one word that explains my meaning in English whereas it would be a whole sentence in German." And it would, of course. He wrote very well in English.

He wasn't really too bad the last two or three days. He was in his room, and the night before he died he was sitting there watching the sunset, and I was with him and Franz [his son] was there, and he said, "Ruth, take my keys and get me a very good bottle of wine out of the cellar."

"What kind would you like?"

"You choose," he said. "We are going to have a very good bottle of wine."

So I went down and got him one of the best wines I knew, for he taught me all I knew about wine, and he taught me an awful lot. He let me choose it for dinner parties. Afterwards, there were only two of us who had keys. And I brought it up with some glasses, and Franz and he and I had some of this wine. Can't remember its name now. But it was one of the best ones, I know, because I knew where they were in the cellar. He seemed to enjoy that very much. And then he went to bed and never got up again. He died the next afternoon about four o'clock. He just went to sleep.

GB: Did he wake the next day?

RB: Yes, but the next afternoon he sort of went to sleep. He did try and tell me something, but I couldn't get it and neither could Franz—about something he saw—but we couldn't get it, then nothing else. But his mouth was very dry, and I got a bit of flannel and some camomile and was wetting it for him, and he seemed to think that was wonderful. And then Franz said, "Don't bother, Ruth, it's too late." But I went on to the very end because his mouth had gone so dry.

GB: What were the last things he said to you?

RB: I cannot remember. I don't think he did say very much that day. The thing that stuck in my mind was that I had to fetch this bottle of wine, and we would have a drink together because he said—he told Fowler—there was one thing he liked about me so much was that I enjoyed food and drink.

Then we had a terrific thunderstorm that night that struck the tree—he died in the afternoon. And that evening when we sat in the salon this big tree was struck—it was terrific, it was. I said to Aggie, "That's something that's gone, it struck something, I am sure."

We didn't know until the next day when we saw all the splinters about everywhere.

And then we had this tremendous funeral service, you know, with about four people giving addresses. And the flowers, from all over the world. I occupied my time seeing faces in a great big arc of roses to keep my mind off it. We had [another] thunderstorm, and it bumped and bumped. Aggie was sitting next to me and she said, "That's Father grumbling."

How he would have hated all that fuss.

*

Ruth Bailey's account depends on her recollections, but having known the principal people she refers to and being familiar with the main events, I believe her descriptions are substantially accurate. However, I cannot confirm or refute the accuracy of the views she says Toni Wolff held about alchemy. Ruth had a superficial interest in Jung's writings and in the significance of symbols, but the details were never of great consequence to her. It was actual people and the details of domestic life and human relationships which really engaged her, and where her recollections are more valuable.

Ruth Bailey was a woman of great warmth, wisdom and integrity, who found her fulfillment in being able to look after Jung and to make it possible for him to spend the last phase of his long and turbulent life in comfort and a creative contentment.

1. See *Jung: His Life and Work* by Barbara Hannah (London: Michael Joseph and New York: G.P. Putnam), p. 347, for details of these dreams, and another one about "a lot of vessels, pottery vases, on the right side of a square place."

2. I think she is referring to *Man and His Symbols* (London: Aldus and Garden City, N.Y.: Doubleday), the first chapter of which was written by Jung in English.

Books of Permanent Jungian Interest

God and the Unconscious *Victor White*

In chapters titled "Psychotherapy and Ethics," "Devils and Complexes," "Revelation and the Unconscious," and "Gnosis, Gnosticism and Faith," Victor White, friend, colleague—and at times sharp critic—of Jung's, explores the relationship between theology and psychology. Includes a foreword by Jung and a new introduction by the poet William Everson, in whose life Victor White played a pivotal role. A classic of Jungian psychology. Indexes. (xxxiii, 245 pp./ISBN 0-88214-503-7)

Incest and Human Love *Robert Stein*

Presents a new and radical model for psychotherapy that values erotic differentiation more than the expansion of ego control. Stein's vision emerges out of a sustained examination of the Incest Taboo, which he regards as the key to psychological development, showing its relevance to work, fantasy, and culture as well as to relationships in marriage, in the family, and between patient and analyst. Uncompromising in its challenge to conventional ego psychology, Jungian conservatism, Freudian reductionism, and to every school of thought that veers away from the animal level of the human soul. (200 pp./ISBN 0-88214-506-1)

Insearch: Psychology and Religion *James Hillman*

Widely used in pastoral counseling and psychotherapeutic training, this book sets out the fundamental principles and attitudes of Jungian psychology in a simple, yet deeply experiential style. "Probably Hillman's most humanly feeling book . . . recommended for dream interpretation and practical examples." Although translated into Dutch, German, Japanese, and Italian, the original edition had been out of print for years. (126 pp./ ISBN 0-88214-501-0)

Animus and Anima *Emma Jung*

Two classic papers on the elemental persons of the psyche. Examines both animus and anima as they appear in behavior, fantasies, dreams, and mythology. Accessible, incisive, and with plenty of practical counsel, this book maps a way toward the union of opposites and the emergence of the Self. Includes a picture of the author. (94 pp./ ISBN 0-88214-301-8)

An Introduction to the Interpretation of Fairytales *Marie-Louise von Franz*

The basic course for a psychological understanding of fairytales. Includes a review of the literature, various theories of fairytales, an examination in detail of one example, and a long chapter on the shadow, animus and anima. Index. (160 pp./ISBN 0-88214-101-5)

HARVEST

Journal for Jungian Studies
published by the Analytical Psychology Club
LONDON

Editor: Joel Ryce-Menuhin

HARVEST No. 32 1986 will be available in early November.
Articles will include:

The Emergence of Religious Symbolism during Analysis	*Andrea Dykes*
The Saving of the Child or the Theft of Time	*Wolfgang Giegerich*
Psychoanalysis and Witness	*David Holt*
Faye Pye Memorial Lecture V: Seasons and Dreams	*Sonja Marjasch*
Gnosis and Psychology	*Professor Dr G Quispel*
A Sonata in the Sandplay - Illustrated	*Joel Ryce-Menuhin*
Shakespeare's *A Midsummer Night's Dream* – A Contribution to the Process of Individuation	*Franz Riklin*
Concern with Death and Eternity	*Paul Seligman*

Also:
Correspondence and Reply to Harvest 1985, edited by David Holt.
Poetry section with contributions from Noel Cobb, Kathleen Raine, Eva Seligman
BOOK REVEWS edited by David Black.
These will include:

The Father ed. by Andrew Samuels	*Giles Clark*
The Cult of the Black Virgin by Ean Begg	*John Costello*
The Vertical Labyrinth by Aldo Carotenuto	*Moira Duckworth*
Myself: My Many Selves by Joseph Redfearn	*Graham Fuller*
The Unconscious in its Empirical Manifestations by C A Meier	*David Holt*
The Analytic Experience by Neville Symington	*Christopher Perry*
Forms of Feeling by Robert Hobson	*Peter Tatham*

ISSN No. 0266-4771

ORDER FORM
To: Harvest Administration, 37 York Street Chambers, York Street, London W1H 1DE, England.
Price per copy US$18 or £8 sterling.

NAME: . . . .

ADDRESS: . . . .

. . . . Mailing List for 1987 issue ☐

Please make cheques payable to Analytical Psychology Club (Harvest).